Environmental Science and Engineering

The ultimate goal of this series is to contribute to the protection of our environment, which calls for both profound research and the ongoing development of solutions and measurements by experts in the field. Accordingly, the series promotes not only a deeper understanding of environmental processes and the evaluation of management strategies, but also design and technology aimed at improving environmental quality. Books focusing on the former are published in the subseries Environmental Science, those focusing on the latter in the subseries Environmental Engineering.

Priyanka Mahajan · Virat Khanna

Graphene in the Environment

Green Solutions for Heavy Metal Contamination

Springer

Priyanka Mahajan
University Centre for Research
and Development
Chandigarh University
Mohali, Punjab, India

Virat Khanna
University Centre for Research
and Development
Chandigarh University
Mohali, Punjab, India

ISSN 1863-5520 ISSN 1863-5539 (electronic)
Environmental Science and Engineering
ISBN 978-3-031-96710-8 ISBN 978-3-031-96708-5 (eBook)
https://doi.org/10.1007/978-3-031-96708-5

This Springer imprint is published by the registered company Springer Nature Switzerland AG
The registered company address is: Gewerbestrasse 11, 6330 Cham, Switzerland

Preface

Both the environment and human health are greatly affected due to the harmful impact of heavy metals, so their remediation always stands as a key challenge in the field of environmental science and nanotechnology. The special qualities of graphene have emerged as a new beginning in the search for effective and long-lasting solutions.

The main focus of this book is to explore the use of graphene in heavy metal cleanup. The introductory chapter contains enough information for understanding graphene, including its synthesis, characterisation, and amazing mechanical, electrical, and thermal properties. The risks of compounds like lead, mercury, cadmium, and arsenic are also explained, along with the dangerous effects of heavy metals. The next chapters go into great detail about the ways by which graphene expertly adsorbs heavy metal ions. The ability of graphene to absorb heavy metals and the extraordinary adsorption characteristics that explain graphene's efficiency in heavy metal removal are well explained. The book also shows how graphene may be dynamically combined with other materials to create potent nanocomposites with improved heavy metal cleanup capabilities. The environmental and safety implications of these ground-breaking solutions are also well given in detail. In addition to discussing the potential environmental issues associated with the discharge of graphene particles, the complexity of graphene dispersion, scalability, and cost-effectiveness are also covered. The book ends by looking towards the future prospects of how graphene will change its role in heavy metal cleaning and continue to advance in the same domain.

This book is an attempt to explore a cutting-edge area of materials science and environmental protection. It connects the extraordinary qualities of graphene and heavy metal contamination in pursuit of a cleaner environment and is a good resource for scholars, academicians, and environmental practitioners by revealing a thorough.

Mohali, India

Priyanka Mahajan
Virat Khanna

Competing Interests The authors have no competing interests to declare that are relevant to the content of this manuscript.

Contents

About the Authors

Dr. Priyanka Mahajan is an Assistant Professor at the University Centre for Research and Development (UCRD), Chandigarh University, Mohali, Punjab. With over nine years of academic and research experience, she brings expertise in plant sciences, particularly in eco-physiology. She holds a Ph.D. in Botany from Panjab University, Chandigarh, where her research focused on mitigating the phytotoxic effects of hexavalent chromium in cereal crops using Beta-pinene, a natural monoterpene. A Gold Medalist in her M.Sc. (Botany), Dr. Mahajan has published 18 high-impact research and review articles, authored five book chapters, and edited two books. She has actively participated in numerous national and international conferences, showcasing her commitment to advancing plant science research and education.

Dr. Virat Khanna is an accomplished academic and researcher with over 15 years of experience in higher education and scientific research. He has authored more than 70 publications in high-impact, peer-reviewed journals indexed in SCI, Scopus, and Web of Science. His research contributions span nanotechnology, materials science, and sustainable development. Dr. Khanna has served as an editor for multiple books and conference proceedings with leading international publishers, including Springer, CRC Press, IGI Global, and Bentham Science. He is actively involved in scholarly publishing and currently serves on the Editorial Board of *Scientific Reports* (Nature Group) and *Discover Nano* (Springer). Additionally, he holds the position of Academic Editor for *Journal of Nanotechnology and Advanced Materials Science and Engineering*, both published by Wiley. As a recognized leader in his field, Dr. Khanna is the Book Series Editor for *Recent Advances in Materials for Nanotechnology*, published by CRC Press (Taylor & Francis). He has also guest-edited several special issues for renowned international journals, including those published by Springer and De Gruyter. Dr. Khanna has delivered numerous invited seminars, keynote addresses, and workshops at prestigious international conferences. His contributions to research and academia have been recognized with multiple best research paper awards.

Chapter 1
Introduction to Graphene and Heavy Metal Contamination

1.1 Introduction

1.1.1 Graphene

Graphene is a carbon-derived nanomaterial (NM) which consists of two-dimensional (2D) layers of a single atom-thick planar sheet embracing sp^2-bonded carbon atoms arranged in a hexagonal honeycomb lattice (Choi et al. 2010; Huang et al. 2015). Alternatively, its structure can be visualised as numerous benzene rings joined together, and the only difference is that the hydrogen atoms are replaced by the carbon atoms (Fig. 1.1a). As graphene holds hybridised sp^2 bonding, it exhibits one orbital and 3 sigma (σ) bonds in perpendicular orientation to the plane (Fig. 1.1b).

Graphene is considered a mother block for all graphitic materials as it possesses a tendency to be wrapped into 0d fullerenes, rolled into 1d carbon nanotubes (CNTS), and stacked into 3d graphite as shown in Fig. 1.2. Owing to the structural network of graphene, viz., covalent bonding of carbon atoms within the plane, associating with three nearby carbon atoms through σbonds and out-of-plane one-bond, graphene material displays unique attributes (Trivedi et al. 2019). Graphene has been labelled as one of the most impactful and advantageous allotropes amongst the whole nano-carbons family due to its magnificent properties such as extremely high specific surface area in the range 2600 m^2/g, outstanding electronic conductivity in the range 2.5×10^5 $cm^2/V/s$, tremendous mechanical strength with Young's modulus of 1 TPa, superior chemical stability, high optical transparency exhibiting nearly 97.4% transmittance observed at 550 nm, and superb thermal conductivity in the range nearly 3000 WmK^{-1} (Gadipelli and Guo 2015; Peigney et al. 2001; Novoselov et al. 2012).

While previous assumptions of graphene being unstable in developing wrapped frameworks, such as fullerenes and CNTS, as it was known to be a vital component of 3d materials and presumed not to occur in a free state, were later found to be false. In 2004, astonishingly free-standing graphene was found, evident by its massless Dirac

P. Mahajan and V. Khanna, *Graphene in the Environment*, Environmental Science and Engineering, https://doi.org/10.1007/978-3-031-96708-5_1

Fig. 1.1 **a** Diagrammatic representation of graphene layer structure (Ghany et al. 2017) and **b** Illustration of out-of-plane π orbitals and in-plane σ bonds, respectively oriented perpendicular and parallel to the plane of the graphene sheets (Mbayachi et al. 2021)

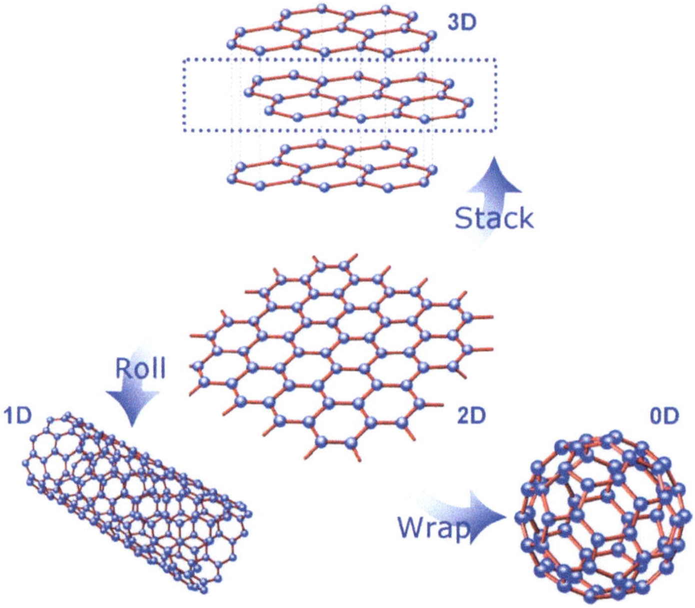

Fig. 1.2 Pictorial view of graphene: the basic building block for other carbon allotropes, graphite (3d), fullerene (0D), and CNT (1D) (Wan et al. 2012)

fermion charge carriers (Zhang et al. 2005; Garg et al. 2014). In this regard, Geim and Novoselov were awarded with Nobel Prize in physics in 2010 for conducting novel research related to graphene material.

Figure 1.3 gives a schematic representation of the timeline of events which have occurred in the history of graphene.

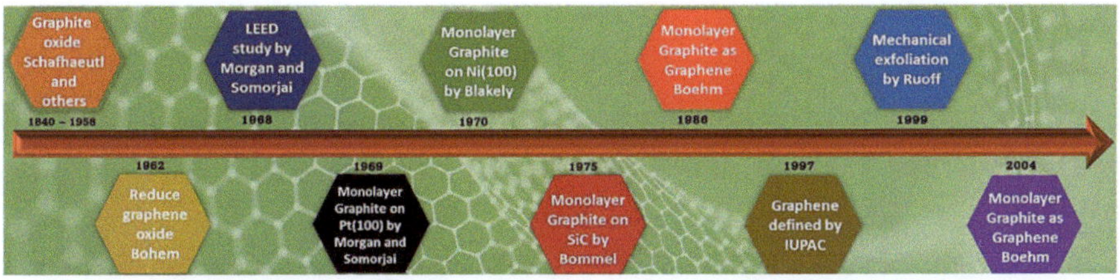

Fig. 1.3 Timeline view illustrating the fundamental milestones in the historical development of graphene (Abbas et al. 2022)

1.1.2 Graphene Synthesis Mechanisms

The two major synthesis approaches for graphene are top-down and bottom-up pathways, as illustrated in Fig. 1.4. In case of top-down techniques, bulk precursor, graphite in the present scenario, is converted into small components for obtaining the product material, viz., graphene in this case (Kumar et al. 2021). The most common top-down techniques for graphene synthesis are plasma etching, liquid-phase exfoliation, laser ablation, electrochemical exfoliation, chemical reduction, and ball milling (Shah et al. 2018; Cai et al. 2018; Torres-Mendieta et al. 2016; Prekodravac et al. 2021; Rabchinskii et al. 2020; Moreno-Fernández et al. 2021). Contrary to this, bottom-up techniques of graphene synthesis involve degradation of carbon-based precursors, including both gaseous and liquid and thereafter developing a hexagonal framework of graphene layers (Yoon and Dong 2020). The various bottom-up approaches include epitaxial growth, thermal pyrolysis, and chemical vapour deposition (CVD) (Gao et al. 2020; Shang et al. 2018; Sun et al. 2021). The most frequently and widely used graphene synthesis techniques are elaborated below.

Chemical exfoliation: In this technique, colloidal suspensions are created via a chemical process, which facilitates the exfoliation of graphite and a graphite intercalating agent into graphene (Syam Sundar et al. 2023). The chemical process involves the isolation of graphite layers either through oxidation or reducing solvent. These reducing or oxidising agents widen the interlayer distance of the graphite layers, thereby diminishing the van der Waals attraction between them (Sahu et al. 2021; Zhang et al. 2010). This graphene synthesis pathway seems to be highly efficient and economic as compared to other methods such as epitaxial growth, Hummer's methods, and micromechanical cleavage.

Mechanical exfoliation: This process of graphene fabrication, though not very prevalent, is yet highly significant for preparing single-layer graphene flakes on a peculiar substrate. This includes a bunch of top-down approaches such as ball milling, sonication, micromechanical cleavage, intercalation, and liquid-based exfoliation.

Micromechanical cleavage: This approach for graphene synthesis is also termed the Scotch tape or peel-off technique, which was used by Geim and Novoselov for the very first instance of graphene synthesis, applying adhesive tape for tearing the graphene layers apart (Fig. 1.5) (Novoselov et al. 2005; Edwards and Coleman

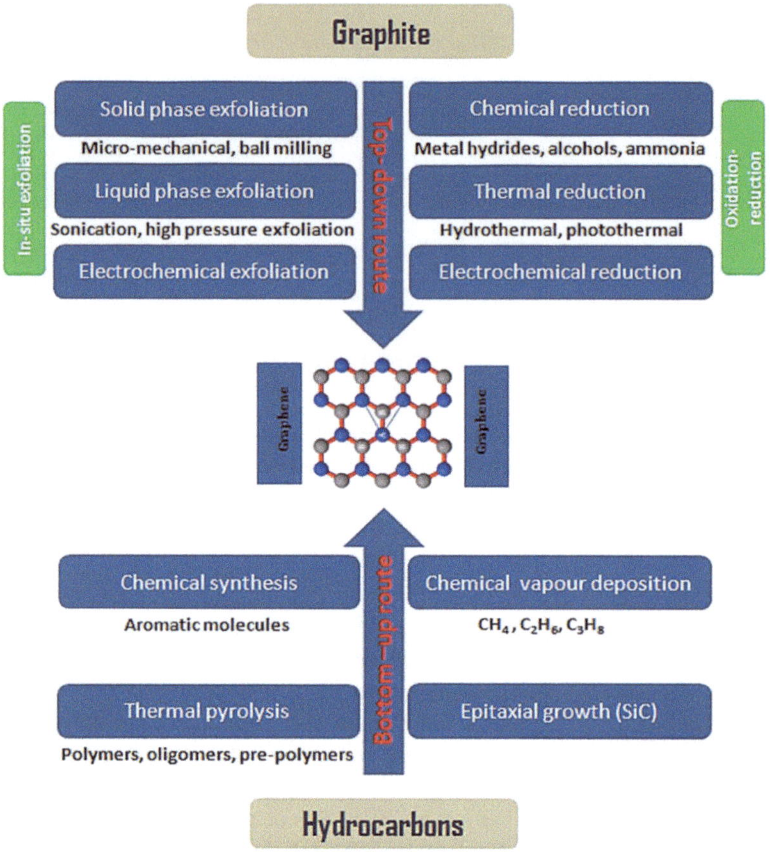

Fig. 1.4 Illustrative presentation of the most commonly involved synthesis techniques for graphene (12)

2013; Noorden 2012). Although several graphene layers get adhered to the tape after peeling off, repeated peeling processes result in the separation of a sufficient number of graphene flakes. The tape is required to be attached to an appropriate substrate, such as acetone, to attain its separation, and at the final peeling step, fresh tape is used for obtaining graphene flakes of variable dimensions (Casiraghi et al. 2007). The obtained graphene flakes are of inferior quality, hence used only for studying their properties rather than applying them for commercial applications (Edwards and Coleman 2013).

Liquid phase exfoliation: For minimising the van der Waals attraction between the graphene layers, graphite is mixed with organic solvent such as N-methyl-2-pyrrolidine (NMP) or aqueous solution before its treatment with external forces, viz., sonication, shear mixing or ball milling, resulting in the generation of graphene flakes (Edward et al. 2023). Although, liquid phase exfoliation technique assists in the production of large-scale and high-quality graphene owing to its facile, eco-friendly,

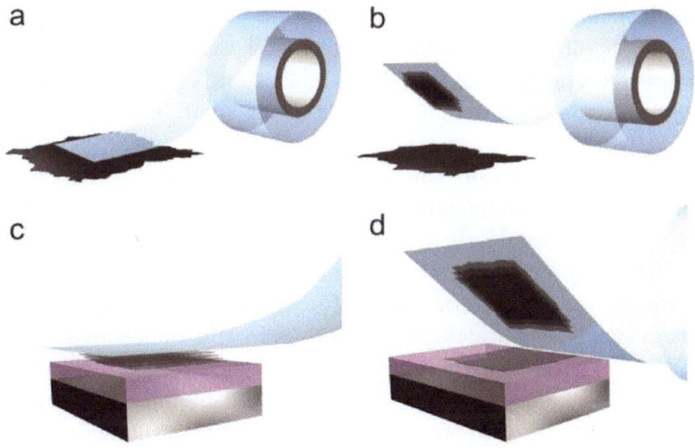

Fig. 1.5 Represents the step-wise schematic process of mechanical exfoliation process **a** the uppermost layer of the adhesive tape are brough in contact of highly ordered pyrolytic graphite (HOPG), facilitating adhering of layered crystal material, **b** tape containing exfoliated layers is pressed on the targeted substrate surface, and **c** & **d** lowermost layer remains attached to the substrate after peeling of the tape (Syam Sundar et al. 2023)

and scalable process, yet it is associated with various limitations such as high cost, non-homogeneous graphene flake dispersion, robust energy consumption, and time ineffectiveness. Hence, its industrial-scale application requires various amendments to overcome the associated drawbacks.

Ball milling: This seems to be a superior technique utilised for synthesising high-grade graphene on a large industrial scale. In this, high speed is applied to break down the graphite in a closed rotating container with steel balls. The obtained graphene flakes are then dispersed in a liquid. It involves the lateral application of shear force, leading to graphitic exfoliation into graphene flakes. Two types of forces act, viz., lateral and vertical. Amongst these two, lateral force is highly demanding as it facilitates the synthesis of wider-sized graphene flakes, whereas vertical force is less prevalent due to its impact on reduced flake dimension and, in some cases, may completely convert the material into amorphous form. Hence, vertical force is generally kept minimum for obtaining high-quality and larger-sized graphene flakes (Knieke et al. 1259; Sumdani et al. 2021).

Sonication: In this, cavitation is created using sound energy (Brennen 2013). Through a sonication bath or probe, the liquid solution is stirred to generate cavitation bubbles, which then result in shockwaves on collapsing. As the shockwaves pass through graphite, it generates tensile stress. The tensile stress results in peeling off graphene layers due to multiple cavities in the graphite (Xu et al. 2018). The efficiency of the sonication technique for synthesising graphene depends upon the configuration of the container manifested in sound waves, and the dispersion volume (Nuvoli et al. 2011). There are a few defects in the graphene synthesised via this technique owing to extreme pressure and cavitation-induced local temperature (Suslick et al. 2011).

High-shear mixing: Graphite with the solution is rotated at a high speed using a rotating blade or rotor–stator within the confined area for generating shear stress which peels off the graphene layers. The synthesis rate of the graphene with rotor–stator mixing depends upon the constituents of the solution as well as the volume (Paton et al. 2014).

Electrochemical exfoliation: This technique for graphene synthesis has attracted a lot of attention in the recent times due to its high efficiency and non-complex process for obtaining bulk graphene (Su et al. 2011). Different graphite forms, including foils, rods, plates, and graphite powder, are used as electrode material in a non-aqueous or aqueous electrolyte and electric current for expanding the electrodes. Depending upon the power supply, electrodes can be positive (anodic) or negative (cathodic) (Liu et al. 2019; Yang et al. 2016). This provides a cleaner and eco-friendly pathway for the synthesis of superior quality graphene from reused graphite, including electrodes from the dumped batteries. The process can be illustrated as submerging the graphite rod in an electrolyte solution, accompanied by the counter electrode. As the current is initiated and passes through the electrodes, the intercalation of the graphite rod with the electrolyte is conducted, leading to effective graphene flake exfoliation. These flakes are then separated via a vacuum filtration process and disbursed in the solvents (Wang et al. 2017).

Chemical vapour deposition (CVD): This technique involves the deposition of material as a thin film onto a substrate using vapour moieties via chemical reactions. This is one of the most prevalent bottom-up techniques for the synthesis of mono-layer or few-layered graphene films. Various complex factors govern the process and involved reactions occurring in the CVD reactor, such as gas feedstock, gas ratio, reactor dimensions, system setup, temperature, growth duration, reactor and gas partial pressure, etc. The very initial thermal CVD process was reported in 1966 for preparing highly crystalline graphite films on nickel (Ni) substrates (Karu and Beer 1966). The process can be explained as the thermal decomposition of carbon-rich sources, usually a hydrocarbon, at an extremely high temperature, and it leads to graphitising of the carbon atoms onto a transition metal substrate, usually Ni or Cu (copper). Copper is used for generating monolayered graphene as it possesses less carbon solubility and controlled catalytic output. Thereafter, with the assistance of support polymers such as polymethylmethacrylate (PMMA), the graphene film is shifted to a transferring agent, viz., silicon or silicon dioxide. With the help of ferric chloride, the Cu substrate is exfoliated for placing PMMA and graphene on a target substrate. The PMMA support is then removed via the use of acetone, and few-layered graphene is obtained (Wang et al. 2017).

Epitaxial growth: It is the process of synthesising graphene layers from silicon carbide (Sic) crystals via thermal decomposition at an extremely high temperature (more than 1000°c) and vacuum. On exposure to such a high temperature, Sic crystals decompose to their components, viz, Si and C. The thermal condition vaporises the Si from the crystal structure, and the left C atoms develop into graphene layers. The obtained graphene exhibits superior quality with minimal wrinkles and ordered crystalline films (Lin et al. 2019). The graphene layers synthesised through this technique possess the additional benefit of no need to transfer the graphene to other substrates and enabling it to produce electronic devices and sensors straightforwardly on the semiconducting or semi-insulating materials (Sic) without any hassle, as compared to other bottom-up approaches such as CVD (Mishra et al. 2016; Beshkova et al. 2016). The major challenge lies with the high temperatures for Sic decomposition and the high cost of Sic wafers, leading to size restrictions of graphene sheets. This delimits the usage of epitaxially synthesised graphene for high-frequency transistors (Lin et al. 1979).

1.1.3 Properties of Graphene

Graphene, owing to its unique physicochemical properties, has gained a lot of attention for environmental applications.

Electronic properties: As mentioned above, carbon atoms in graphene are arranged hexagonally and create triangular mesh in connection with the two atoms, resulting in a honeycomb framework. This provides graphene with remarkable electronic attributes, with conductivity ranging up to 106 S/m and sheet resistance of 31 Ω/sq (Kim et al. 2009). It can bear current densities much higher than Cu, reaching up to six orders of magnitude (Geim 1979). Owing to ambipolar electric field impact, graphene can spontaneously and repeatedly switch between holes and electrons with a dosage of 10^{13} cm^{-2}. Graphene exhibits semi-metallic properties due to minute overlapping between the valence band (VB) and conduction band (CB). These VB and CB possess a conical structure, hence they cut off at the Dirac point (Karthik et al. 2021). These outstanding electronic attributes of graphene are obtained under ideal conditions (Bolotin et al. 2008). However, there are various factors on which the electronic properties of graphene depend, such as defects, impurities, graphene layers, functional groups, etc. (Yang et al. 2010; Sreeprasad and Berry 2013). Nevertheless, the magnificent electronic attributes of graphene make it a highly demanding material for environmental applications (Yang et al. 2010; Zhao et al. 2012).

Mechanical properties: The exceptional mechanical attributes of graphene make it a suitable material both as an individual and as a supporting material for developing nanocomposites. The mechanical stability is provided by the sp^2 bonds of the hexagonal lattice of graphene, as well as the ability to reduce the number of in-plane deformations (Papageorgiou et al. 2017). It is observed that the monolayer graphene, which is free from any defect, is found to be the strongest material as the intrinsic strength is reported to be 42 N m^{-1} by Hone and colleagues, equalising to the 130 GPa intrinsic strength (Lee et al. 1979). Besides, two highly remarkable mechanical properties of graphene are wrinkling and crumpling, which are extremely important in the making of complicated nanomechanical frameworks. The wrinkled and crumpled graphene has undergone elaborate research in Buehler's laboratory via analysing its surface modifications, substrate, and folding as well as unfolding of graphene in a controlled way (Al-Mulla et al. 2015; Qin et al. 2014). Further, the internal stiffness and strength of defective graphene were also studied by introducing defects in the pristine graphene through modified oxygen plasma, as well as the atomic force microscope nanoindentation technique. The defect obtained via these techniques was sp^3, and it was observed that both stiffness and strength of defects-induced graphene remained the same, whereas only 14% lower breaking strength was seen as compared to pristine graphene.

Surface area: Graphene possesses an extremely high specific surface area, as each atom present in the single layer of the graphene sheet is in contact with the environment from both sides (Sanchez et al. 2012). Amongst almost all the nanomaterials, graphene exhibits the highest specific surface area, ranging up to 2,630 $m^2\ g^{-1}$. Owing to this high surface area, graphene represents the ideal candidate for carrying out the surface reactions and adsorption of pollutants, thereby highly suitable for environmental remediation processes. Besides, this property of graphene gives extensive opportunity for developing nanocomposites as it provides suitable support to attach other nanomaterials and chemical moieties (Compton and Nguyen 2010).

Surface functional groups: The oxidised form of the graphene, viz., graphene oxide (GO), provides a cost-effective way of utilising graphite and confers it with additional benefits of highly dense oxygen-containing surface functional groups such as hydroxyl, carboxyl, epoxy, and carbonyl in the graphene carbon lattice. Synthesising GO through chemical oxidation of graphite, followed by exfoliation via ultrasonication, is a budget-friendly technique (Dreyer et al. 2010). Besides, graphite oxidation leads to widening of the interlayer space from 0.34 to 0.65 nm, thereby reducing the amount of energy needed to move graphene layers apart (Hontoria-Lucas et al. 1995). Graphene oxide is gaining immense importance owing to its diverse attributes, suitable for various environmental applications. The most attractive features of GO are its ability to develop chemical functionalization and its high surface chemical reactivity (Marchi et al. 2018). The covalent oxygenated groups attached to the GO surface provide an area for GO modifications, which further work for tuning its physicochemical properties. The large number of oxygen moieties on the GO surface is of immense importance in environmental remediation applications as they provide important sites for interacting and adsorbing metal ions from the environment (Zhao et al. 2011a).

1.1.4 Characterisation of Graphene

Characterisation of graphene provides essential information regarding the properties, morphology, layers, and defects utilising microscopic and spectroscopic measurements (Chen et al. 2017; Li et al. 2009).

X-ray diffraction (XRD) analysis: This technique is applied to reveal the phase of the material obtained via cell dimension units (Seekaew et al. 2014). The XRD analysis is generally performed to elucidate the occurrence of metallic components in graphene structures. Besides, a diffraction pattern is created when X-rays interact with the crystalline phase of the graphite (Yousefi et al. 2019). Bragg's equation forms the basis of this technique, which highlights the simple relation existing between the

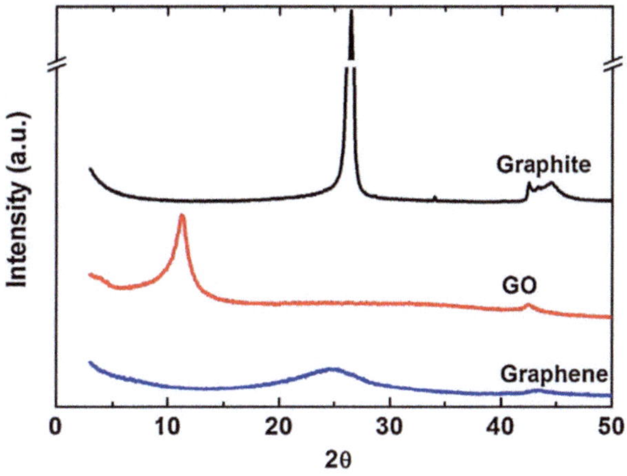

Fig. 1.6 Graphical view of XRD patterns of graphite, graphene oxide and graphene (Johra et al. 2014)

x-ray wavelength (λλ), the crystal's interplanar spacing (d), and angle of reflection (θ) as given in the following equation: $n\lambda = 2d \sin\theta$, where n is the order of lattice planes. This law states that when an x-ray incident on a surface of a crystal framework at an angle of incidence, it is then scattered back with a similar angle of scattering for facilitating the constructive interference, the path difference is equal to a whole number multiple of the wavelength. Further, it was noticed that X-rays can only be reflected at particular angles, unlike ordinary light. These angles are calculated based on the distance between the planes in the crystal as well as the wavelength of the x-rays (The reflection of X-rays by crystals 1913). As shown in Fig. 1.6, a high diffracted and intense peak of pristine graphite is observed at 26.6°. This peak shifts to 13.3°, which evidence the presence of oxygen in the graphitic structure. After the synthesis process, the peak has disappeared, indicating the completion of graphene synthesis (Zhang et al. 2010).

Raman Analysis: The graphene and GO can be characterised via Raman spectra, which are governed by the carbon–carbon double bonds and conjugated bonds, generating highly intense peaks (Fig. 1.7) (Johra et al. 2014). In case of GO, the Raman spectra are characterised by a G band at around 1605 cm^{-1} corresponding to sp^2 C atoms' E_{2g} phonon and representing the D band at 1353 cm^{-1}, indicating the breathing phase of $_K$-point phonons of A_{1G} symmetry. Further, a change in the G band of graphene was observed as compared to the GO band's position, which was found at 1600 cm^{-1}. Moreover, defects like grain boundaries, vacancies, and amorphous C moieties, D band disorder are also introduced (Syam Sundar et al. 2023).

Scanning electron microscopy (SEM) and field emission SEM (FE-SEM): This is used to generate the morphological image of the sample through scanning its surface. This is based upon accelerating the high-energy electron beam through the electromagnetic field, and this beam is focused on the sample surface with the aid of a condenser lens system. After interacting with the atoms present on the sample surface, these electrons are emitted back, also termed as secondary electrons, which are then detected by the electron microscope (Goldstein et al. 2018). The topography

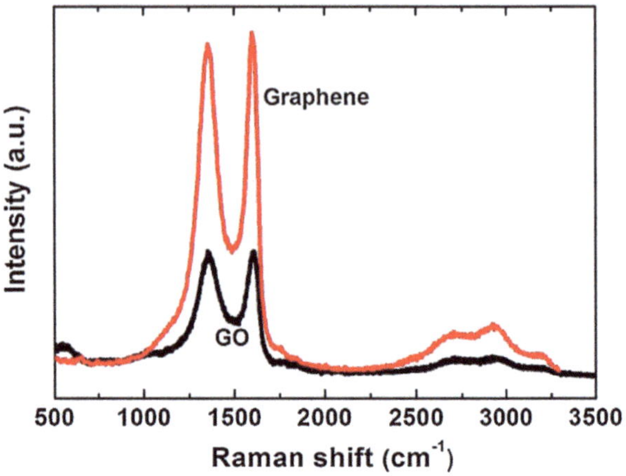

Fig. 1.7 Represents the Raman spectra of GO and graphene (Johra et al. 2014)

of the sample decides the number of secondary electrons. Thereafter, an image is formed exhibiting the sample's topography. Various factors, viz., how intensely the electron beam is focused on the sample surface, how it is dispersed, and how much area it covers, determine the resolution of SEM. Besides, SEM fitted with a field emission cathode in its electron gun generates narrower probing beams both at high as well as low electron energies. This results in attaining superior spatial resolution with minimal sample damage and charging. Thus, the field emitter electron gun is meant to produce a stable and elaborate current in a narrow beam. The images thus developed are referred to as field emission SEM (FE-SEM) images. These micrographs exhibit high resolution with less electrostatic distortion as high as 3 to 6 times than conventional SEM images (Abraham et al. 2020).

Figure 1.8 represents the SEM images of polyethene terephthalate (PET) encapsulated graphene sheets at 2 nm. It is observed that interaction exists between graphene and PET leads to enhanced thickness of the graphene sheets which increased to 50 nm from 1.57 nm. Owing to the enlarged surface area of the graphene nanosheets, the interface area between graphene and PET was enhanced significantly, resulting in large number of tunnelling sites important for electron transport (Zhang et al. 2010).

Figure 1.9 gives an overview of FE-SEM images of graphite, nanographite, multilayered graphene nanoplatelets, and wrinkled-derived graphene from GO (Kaniyoor and Ramaprabhu 2012).

Transmission electron microscopy (TEM): This involves the transmission of a focused beam of electrons through an extremely thin sample film. While transmitting through the sample, the beam interacts with its structure, and based upon the information received from the transmitted electrons, an image is generated exhibiting the sample's topography (Robinson 1986). The TEM-derived images possess ultra-high resolution as well as details about crystallinity, morphology, composition, besides topography, unlike SEM generated images. Figure 1.10 displays the TEM images

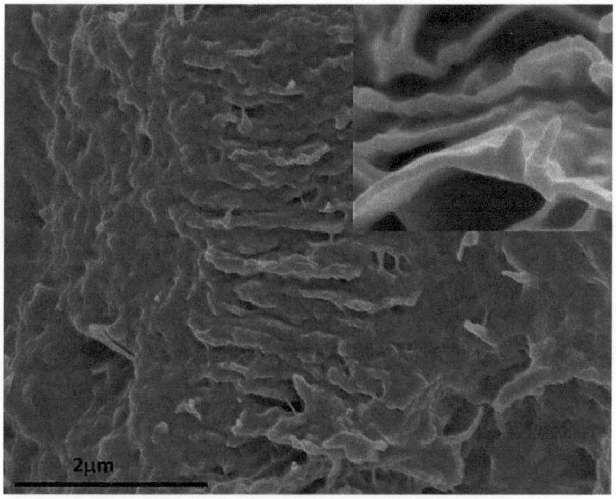

Fig. 1.8 Illustration of scanning electron micrographs of a fractured surface of the PET nanocomposite involving 1.2 vol.% of graphene. Inset represents the highly magnified micrograph of the same nanocomposite (Zhang et al. 2010)

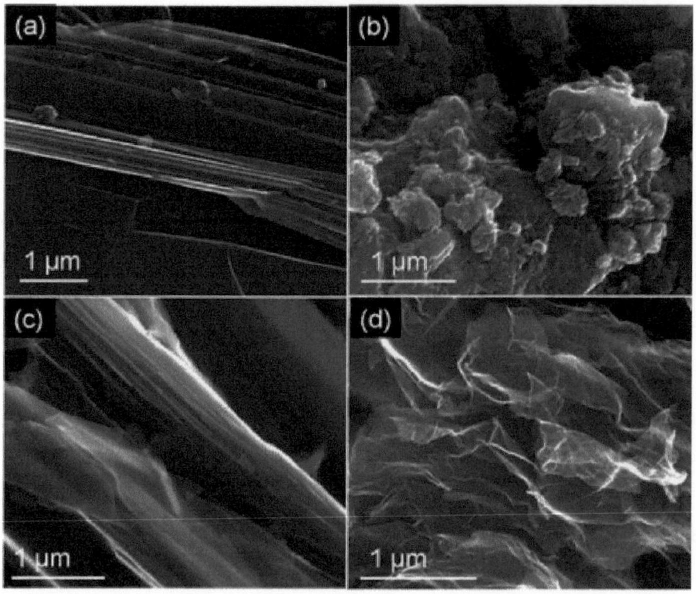

Fig. 1.9 Represents the FE-SEM images of **a** graphite, **b** nano-graphite (nGr), **c** multi-layered graphene nanoplatelets (GNP) and **d** wrinkled derived graphene from GO (Kaniyoor and Ramaprabhu 2012)

of three types of GO nanosheets synthesised via different pathways. All the electron micrographs exhibit proper folding, confirming the presence of oxygen-derived functional groups in their structure (Al-Gaashani et al. 2019).

Fourier transform infrared spectroscopy (FTIR): In this, the infrared (IR) beam is broken into two parts with an internally placed beam splitter inside the interferometer.

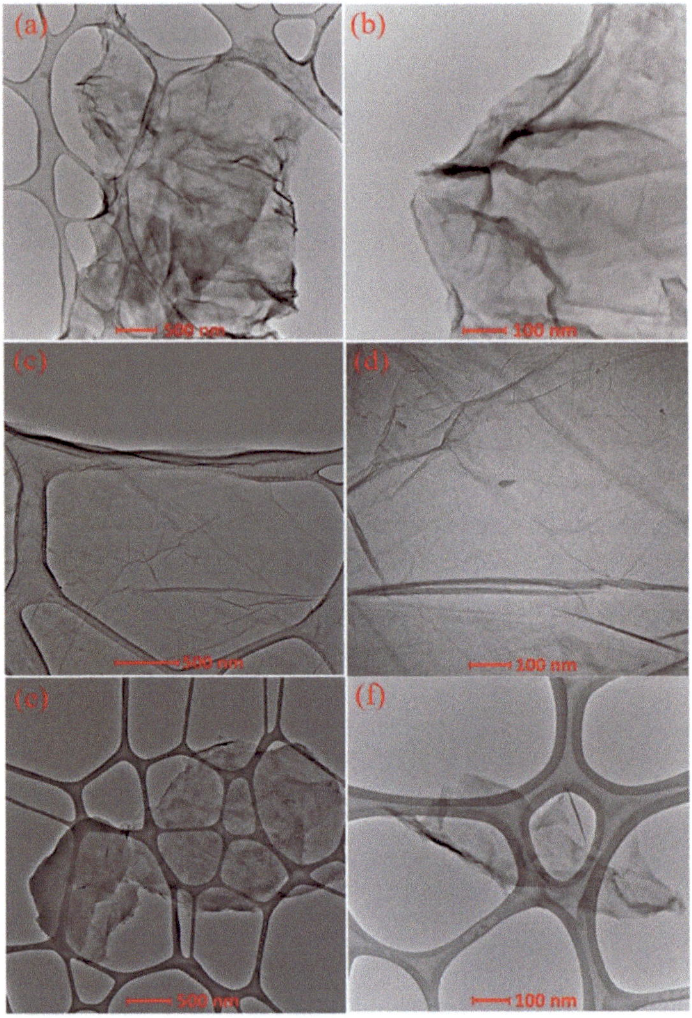

Fig. 1.10 Pictorial view of transmission electron micrographs of GO synthesized via different techniques (Al-Gaashani et al. 2019)

The fixed mirror thus reflects the first part of the beam, and a moving mirror reflects the second part of the beam, which is placed in a perpendicular direction to the fixed mirror. Both parts of the reflected IR beam recombine at the beam splitter and interfere, leading to the generation of an interferogram. Inference signals are created owing to the difference in the optical path by interferogram, get transformed to a spectrum of transmittance or absorbance versus frequency or wavenumber by Fourier transform (Griffiths 1983).

As shown in Fig. 1.11, the FTIR spectra are used to explore the functional groups attached to the GO surface. An intense peak at 3425 cm^{-1} indicates the

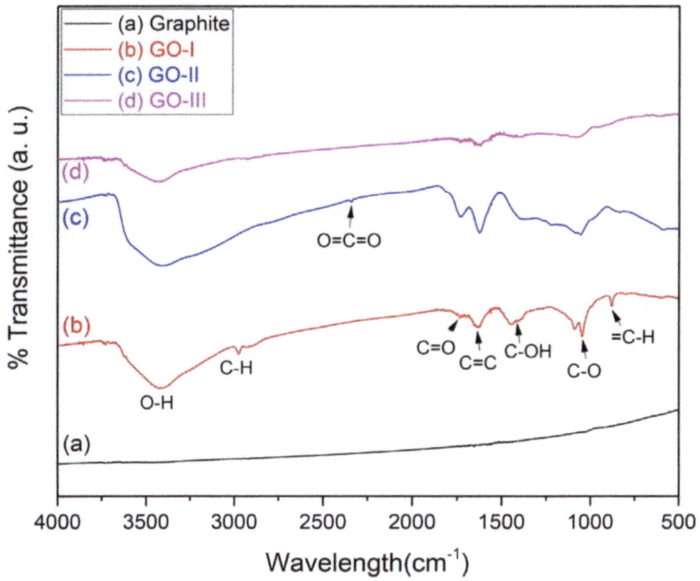

Fig. 1.11 Represents FT-IR spectra of **a** graphite, **b** graphene oxide (GO-I) formed using an acid mixture of H_2SO_4, H_3PO_4, and HNO_3 respectively in a ratio of 70:20:10, **c** graphene oxide (GO-II) synthesized with an acid mixture of H_2SO_4, and H_3PO_4 in a ratio of 90:10, respectively, and **d** graphene oxide (GO-III) developed via ultrasonic bath (Al-Gaashani et al. 2019)

O–H stretching of the COOH group, and the peak of 2972 cm^{-1} represents C–H stretching. The carboxyl (C = O) and aromatic (C = C) stretching groups have been found at peaks of nearly 1735 and 1635 cm^{-1}, respectively. Further, an intense peak at 800 cm^{-1} represents the = C–H bond, whereas wide peaks at 1450 cm^{-1} are owed to O–H deformations in C–OH groups (Konwar et al. 2016). Besides, it is observed that at peak of nearly 2340 cm^{-1}, where the absorption band is centred, could be attributed to atmospheric carbon dioxide (O = C = O).

X-ray photoelectron spectroscopy (XPS): X-rays are used for ionising the innermost core electrons owing to their extremely high energy radiation. The C 1 s XPS spectra of GO and graphene are represented in Fig. 1.12. There are three visible peaks in the case of the C 1 s spectrum of GO, viz., 284.6, 286.6, and 288.5 eV, which correspond to sp^2 C, epoxide, and carboxyl functional moieties (Park and Ruoff 2009). In case of graphene, an extra peak at 287.5 eV was also observed, related to the C = O bond. However, on hydrothermal reduction, the XPS spectrum for graphene exhibited a very intense peak at 284.6 eV, and the intensity was measured to be double that of GO. Whereas, the intensity for sp^2 C, epoxide, and carboxyl functional moieties in this case decreased significantly compared to GO, indicating more percentage of C = C bonds over to C–O bonds, evidencing GO reduction. Moreover, there were no traces of any other elements besides C and O, hence, no impurities were seen (Johra et al. 2014).

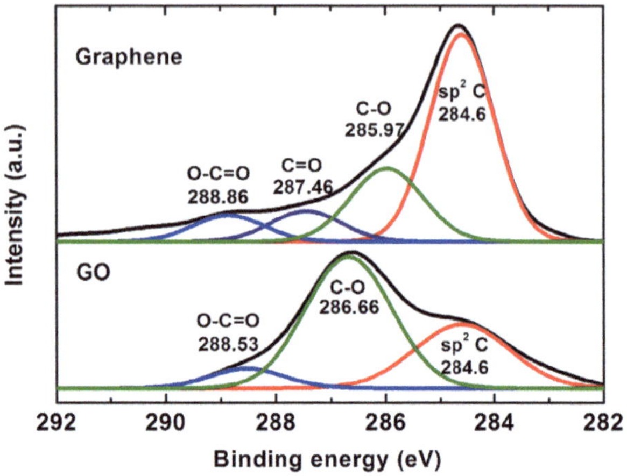

Fig. 1.12 Represents the XPS spectra of GO and graphene (Johra et al. 2014)

1.2 Overview of Heavy Metals and Their Environmental Impact

1.2.1 Heavy Metals

Heavy metals (HMS) are distinguished from other metals in possessing higher atomic weight and density (at least five times more than that of water, > 5 g/cm^3) (Phaenark et al. 2024). However, most appropriately HMs refer to metallic elements and metalloids obnoxious for humans and environment as few lighter metals including arsenic (As), selenium (Se), and aluminium (Al) are categorized under HMs due to their toxic nature, whereas higher weight metals such as gold (Au) is non-toxic, thus not categorized as HMs (Briffa et al. 2020).

Heavy metals are mainly categorised into essential and non-essential metals depending on their role in biological functionalities. Essential metals refer to those elements which in minute amount drive various biochemical processes in the body such as copper (Cu), cobalt (Co), iron (Fe), manganese (Mn), molybdenum (Mo), zinc (Zn), selenium (Se), chromium (Cr), nickel (Ni) etc. Whereas, non-essential metals are of no biological usage and are not required in the body due to zero nutritional importance (Ali and Khan 2018). The other category of HMS is less toxic and highly toxic metals, as mentioned in Fig. 1.13.

Although HMS occur naturally in the environment, including volcanic eruptions, forest fires, soil debris, and sea salt, the foremost responsible factors for tremendously increasing environmental pollution are anthropogenic activities. The major human-related activities causing HM toxicity in the environment include pollutant discharge from vehicular sources, improper disposal of industrial effluents, construction activity, using lead (Pb) as a fuel antiknock agent, and sewage discharge. Moreover, various industries linked with pharmaceuticals, paper and pulp production, chlorine and caustic soda synthesis, and farming introduce an ample amount of

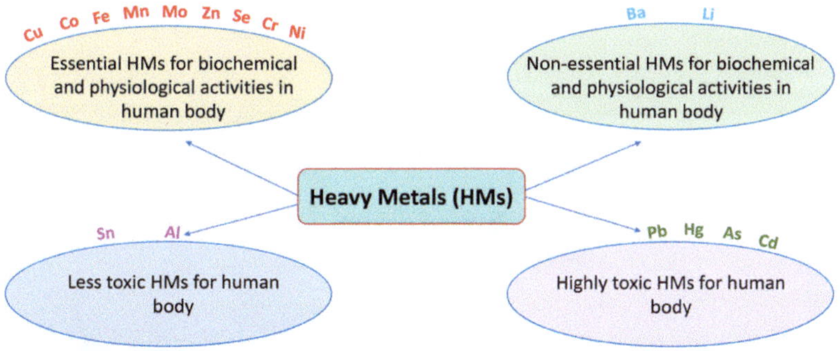

Fig. 1.13 Represents the segregation of heavy metals based upon their impact on the human body (Filipoiu et al. 2022)

mercury (Hg) into the environment. Besides, cadmium (Cd) is released from electroplating industries, which play a major role in developing batteries, metal coatings, and textiles. The extent of toxicity of any HM depends upon its concentration and exposure time. Highly toxic HMs mainly include lead (Pb), mercury (Hg), arsenic (As), cadmium (Cd), chromium (Cr), zinc (Zn), copper (Cu), nickel (Ni), whereas less toxic HMs include tin (Sn), aluminium (Al) (Fulke et al. 2024).

1.2.2 Heavy Metals' Toxic Impacts on the Environment

The HMS discharged either naturally or anthropogenically enter the terrestrial and aquatic ecosystems, and their accumulation beyond permissible limits leads to their contamination. Due to their highly persistent nature, HMS either accumulate in the biota or percolate to the groundwater. Thereafter, consumption of HM-contaminated water leads to serious health hazards in animals and humans as well as developmental issues in plants (Fig. 1.14) (Jomova et al. 2025).

Aluminium (Al): It is the 13th element of the periodic table and is the 3rd most abundant element in the Earth's crust. As a metal atom, Al is zerovalent, however, there are three oxidation states (+1, +2, and +3) of Al in compound form. The major resources and processes of Al-based compounds are photography, oil refineries, and buildings. The compounds are silicates, sulfates, chlorides, nitrates, and hydroxides (Igbokwe et al. 2019). The Al is taken up by the human body through inhalation, dermal contact, and ingestion. Moreover, industrial discharge containing Al-based compounds from coal power plants, metal refining industries, and steel foundries ultimately enters the water bodies, thereby threatening the aquatic life and other living beings dependent upon water for their existence (Gensemer and Playle 1999).

Cadmium (Cd): It is the 48th element of the periodic table and resembles mercury (Hg) and zinc (Zn) in their physicochemical attributes. The most stable oxidation state of Cd is +2. Nearly 13,000 tons of Cd are generated from Ni–Cd batteries, colour pigments, chemical stabilisers, metal coatings, and alloys, worldwide. Besides

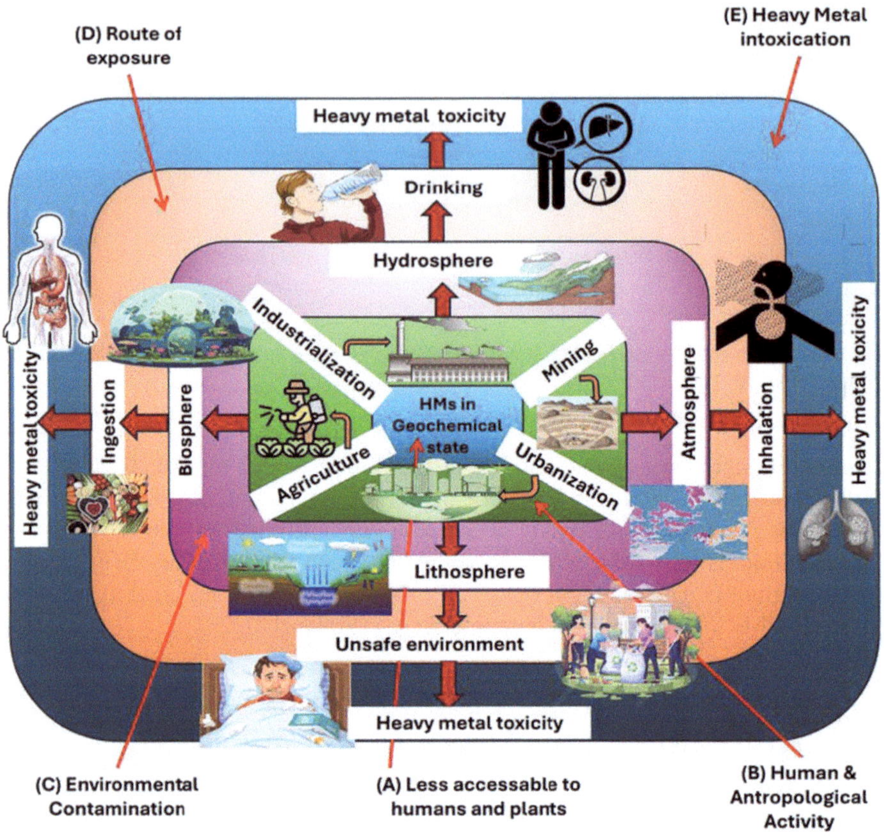

Fig. 1.14 Schematic view of heavy metal ions sources, their interaction, and obnoxious effects on environment and human health (Seid Ali et al. 2024)

volcanic activity and rock weathering, Cd in the environment enters through various human activities, viz., mining and processing of Cd-containing sulfide ores of Cu, Zn, and Pb. The major entry routes of Cd in the human body are the gastrointestinal tract and lungs (Valko et al. 2005). The major sources of water contamination through Cd involve water bodies occurring in the vicinity of hazardous waste sites, landfills, battery recycling areas, mining activity, surplus fertilisers, combustion of fossil fuels, waste incineration, etc. High accumulation of Cd has been observed in the tissues of aquatic organisms (He et al. 2023). Besides, the toxic impact on the seed germination of plants exposed to Cd-loaded P fertilisers has also been reported (Jomova et al. 2025).

Arsenic (As): It is the 33rd element of the periodic table and exists in the oxidation states of -3, 0, +3, and +5. Arsenic exhibits both metallic and non-metallic properties and exists in three allotropic and two organic and inorganic states (Exley 2013). In natural existence, as occurs in igneous rocks, ores, and sediments. The most crucial source of As is red ore arsenopyrite, which contains Fe and S, besides As. In water and soil, As occurs abundantly in inorganic arsenates and arsenites, whereas in water, the predominant form of arsenic is arsenic trioxide. The industrial pool for As includes glass processing, ammunition, paper, textiles, wood preservatives, etc. Besides, as is found in cosmetic products such as colour pigments, eye shadows, which are carriers for their entry into the human body via cutaneous penetration, thereafter into the circulatory system. The major organic sources of As are rice, mushrooms, poultry, and seafood (Jones 2007). The ingestion of As beyond permissible limits leads to carcinogenic impacts.

Mercury (Hg): It is the 80th element of the periodic table and exists in the oxidation states of +1 and +2. Very few compounds of Hg are available with an oxidation state of +1, whereas diatomic Hg is the most stable and most available form. At room temperature, it is found in liquid form and exhibits minute volatility, however, its volatile nature enhances with increasing temperature (Gaffney and Marley 2014). The Hg exists abundantly as mercuric chloride ($HgCl_2$). This compound of Hg has wide-scale applications as a fungicide in agriculture, a catalyst for the production of vinyl chloride, and is available as a precursor for various Hg-based compounds. Another Hg-based compound, viz., mercury sulfide (Hgs), plays a crucial role as a pigment for rubber, plastics, and paints. Mercury resides in elemental, organic, and inorganic states in the environment and has a drastic negative impact on animals and human beings with its prolonged exposure (Fitzgerald and Clarkson 1991). The water, air, and soil contain inorganic salts of Hg in association with S and Cl. In natural conditions, Hg exhibit a lot of physical and chemical transformations, and during this passage, it interacts with organic molecules and generates organo-Hg compounds such as highly toxic methylmercury, which attacks the nervous system. The most common sources of methylmercury for human exposure include the consumption of fish and shellfish.

Lead (Pb): It is the 82nd element of the periodic table, which exists in the oxidation states of +2 and +4, and amongst which the +2 oxidation state is more stable. It is a highly obnoxious element for the environment, animals, and human beings. This element is highly persistent in the environment and accumulates in sediments and soil via air absorption, mining activities, erosion, and waste discharge in the water bodies (Wani et al. 2015). The most common sources of Pb are pipes, gases, plumbing materials, batteries, glasses, ammunition, and cosmetics (Recycling used lead-acid batteries: health considerations). Besides these household resources, the industrial discharge of Pb occurs via the smelting process, mining activity, fossil fuels, etc. As Pb particles take longer to get settled, inhalation is one of the most common routes for its intoxication in human beings. Lead is particularly obnoxious for children and pregnant women as it has a more lethal impact on the developing organs.

Chromium (Cr): It is the 24th element of the periodic table and the 7th most abundant element of the Earth's crust (Singh et al. 2013). Though Cr exists in various oxidation states ranging from –2 to +6, trivalent (+3) and hexavalent (+6) forms are the most stable ones. Amongst these two forms, hexavalent Cr is the most toxic one (Mahajan et al. 2016). The major anthropogenic source of Cr in the environment, besides natural emissions from rock, soil, volcanic dusts, plants, and animals, is industrial discharge. The major industries are leather and wood preservation, metal plating, timber processing, electroplating, production of dyes, paints, and varnishes, oxidative dyeing, tobacco smoke, paper and pulp industries, etc. (Mahajan et al. 2021; Mahajan and Khanna 2025). Chromium-induced environmental contamination has become a major concern worldwide. The Cr enters the human body mainly via inhalation, dermal contact, and oral passage, and its accumulation beyond permissible limits leads to many abnormalities such as pulmonary cancer, skin ulcers, DNA lesions, and oxidative damage. Chromium-induced phytotoxicity has a drastic impact on plant growth and development as well as seed germination (Mahajan et al. 2019).

1.3 Nanomaterials (NMS) and Heavy Metals' Remediation

Metal oxide NMS: These NMS exhibit high potential in the remediation of toxic metals from wastewater (Mudzielwana et al. 2019). Nanodimensional metal oxides are excellent adsorbents due to their extremely high surface area and suitable pore size and structure. Besides, the ability for structural modifications confers it with extra benefits for developing much superior adsorbents out of them (Borji et al. 2020). They include oxides of manganese, zinc, titanium, iron, aluminium, cerium, magnesium, zirconium, etc. Zinc oxide (Zno) NPs, owing to their elaborate surface area, low budget, and magnificent removal tendency, have gained tremendous importance for adsorptive removal of HMS. A study reported the removal of Zn, Cd and Hg using Zno NPs and found the removal efficiency, respectively, in the range of 357, 387, and 714 mg/g obtained via the Langmuir model, indicating the outstanding adsorptive efficiency of Zno (Sheela et al. 2012). Further, a study reported adsorption of mixed metal solutions comprising Cu, Cd, Ni, and Pb by magnesium oxide (Mgo), aluminium oxide (Al_2o_3), and titanium oxide (Tio_2) in the range of 594.9, 114.6, and 49.4 mg/g (Mahdavi et al. 2013). The Mgo exhibited both adsorption and precipitation mechanisms for HM removal, whereas the other two followed only the process of adsorption. It is observed that the HM removal efficiency of metal oxides is dependent upon the type of HM to be adsorbed as well as the functional groups present on the metal oxides. Iron oxide NPs have gained a lot of popularity in the removal of HM ions owing to their non-toxic nature, magnetic attributes, and ubiquitous occurrence (Silva et al. 2025).

Silica nanomaterials: Silica-derived NMS are of prime importance in adsorption of HMS as they exhibit valuable adsorbent characteristics, viz., elaborated surface area, adaptable pore dimensions, and tunable surfaces (Nayl et al. 2022). The outstanding catalytic and adsorptive capacity of these NMS has attracted environmentalists for the clean-up of air pollution and pollutant remediation in their gaseous form (Guerra et al. 2018). Various surface alterations, such as the introduction of hydroxyl groups, further enhance its wetting and gas adsorption abilities. Previous reports reveal that the amine groups on the surface of mesoporous silica mediate the remediation of hydrogen sulphide and carbon dioxide from the natural gas (Huang et al. 2003). Besides, the air-suspended Pb metal is posing a hazard to the environmental quality and human health. Yang and co-workers fabricated silica NPs suitable for capturing the Pb from the air particles, thus addressing the issue of Pb-induced environmental contamination (Yang et al. 2013).

Polymer-based NMS: Polymers play a significant role in the identification and remediation of diverse pollutants, including toxic chemicals (HMS, nitrates, etc.), obnoxious gases (nitrogen oxides, sulfur dioxide, carbon monoxide, etc.), and organic pollutants (volatile organic compounds, aromatic and aliphatic hydrocarbons, pharmaceuticals, etc.). Hence, these polymeric substrates are often integrated with the NPs to enhance their contaminant removal abilities via aggravating their stability and physico-chemical attributes (Guerra et al. 2018). Various polymers, such as poly (amidoamine) or dendrimers (PAMAM), have been reported to treat wastewater comprising metal contaminants such as Cu ions. The surface functional moieties, viz., carboxylates, amines, and hydroxamates- containing dendritic nanopolymers, exhibit high efficiency in encapsulation of a large number of water species such as cationic species Cu, Fe, Ni, Zn, Ag, Au, U, etc. (Diallo et al. 2005). These materials are excellent chelators and filters for the immobilisation of HMIS, thereby playing an important role in water purification.

Silver and iron NPs: Silver (Ag) NPs represent the aggregate of metallic Ag atoms exhibiting a significantly high specific surface area. Besides possessing distinctive physicochemical attributes, Ag NPs are known for their antibacterial activities, subsiding bacterial, viral, and fungal growth (Bruna et al. 2021). They represent the unique ability to capture and scatter the light particles owing to their surface plasmon resonance, conferring suitability for sensing, optical, and imaging devices. Besides antibacterial activity, Ag NPs are referred to as outstanding adsorptive materials. For instance, Ag NPs derived from the cyanobacteria and microalgae extracts display a high ability for the removal of Fe and Mn ions from the aqueous media due to their large surface area. Likewise, iron (Fe) NPs offer high potential for the removal of HMIS due to their enriched surface area-to-volume ratio, enhancing their reactivity as well as adsorption ability. The major phenomenon involved in the removal by Fe NPs is adsorption mediated by electrostatic interaction occurring between HMIS and active sites held by hydroxyl and Fe oxide moieties (Aragaw et al. 2021). The highly obnoxious hexavalent Cr is converted to less toxic trivalent Cr by these NPs via the transfer of electrons on the surface of Fe-based adsorbents. Iron NPs have shown significant output for the removal of Cd, Ni, As, and Cr ions across different ecosystems (Feisal et al. 2024).

Carbon-based NMs: Owing to the structural framework and composition as well as tunable hybridized states, carbon NMs possess distinctive chemical, physical, and electronic characteristics (Ren et al. 2011). The scalable hybridized can form diverse structural configurations such as carbon nanotubes (CNTs), fullerenes, and graphene, when compared with metal oxide NMs (Mauter and Elimelech 2008).

Carbon nanotubes (CNTS): They represent the class of carbon materials in which the graphitic sheets are rolled to create a cylindrical structure in which carbon atoms are held together by covalent bonds (Li and Pandey 2015). Owing to a unique structural framework and physico-chemical properties, CNTS exhibit superior adsorption capacity for the removal of HMS (Ihsanullah et al. 2016). Carbon nanotubes have

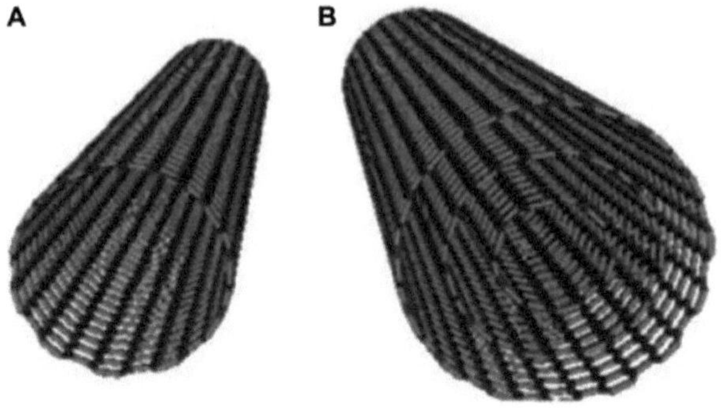

Fig. 1.15 Illustrations of **a** SWCNT and **b** MWCNT (Al Moustafa et al. 2016)

been classified into single-walled CNT (SWCNT) and multi-walled CNT (MWCNT) (Fig. 1.15). Both classes of CNTS exhibit unique characteristics, including rich surface area, low mass density, and large pores contributing towards their high adsorption capacity compared to conventional adsorption materials such as clays, biosorbents, and activated C (Borji et al. 2020). Heavy metal's adsorption via CNTS has been reported by various studies, viz., Cd adsorption, Ni adsorption, Cu adsorption, and Pb and Cr adsorption (Alagappan et al. 2017; Farghali et al. 2017; Hayati et al. 2017; Mubarak et al. 2016).

Depending upon the chemical nature of the contaminant, CNT surfaces are functionalized at different interfaces such as tips and sidewalls, particularly for efficient output in case of wastewater treatment. Different mechanism are involved to bring about the modifications of CNT properties towards hydrophilic contaminants. Functionalized CNTs and its adsorption capacity has been mentioned in the Table 1.1.

Graphene-based NMs: As mentioned above, graphene and its derivatives possess unique physicochemical characteristics including extremely high surface area, tendency to build surface active sites with high tendency of functionalization and modification, and characteristic π-electron system, facilitating its efficiency for potential adsorbents of HMIs as well as many other contaminant groups such as dyes, pharmaceuticals, and antibiotics etc. (Silva et al. 2025; Krishna et al. 2023). Graphene and its derivatives have gained a lot of attention in the field of HM remediation, especially graphene oxide (GO). The extraordinary properties of GO, viz., high surface-to-volume ratio and innumerable oxygen-derived surface functionalization such as hydroxyl, carbonyl, and carboxyl groups. Besides, its excellent hydrophilicity facilitates its enhanced solubility in aqueous media, ensuring maximum interaction between the GO and HMIS (Ahmad et al. 2020). Various graphene-based materials involved in the adsorption of different HMIS have been mentioned in Table 1.2.

Table 1.1 Represents functionalized CNTS for the removal of different HMIS along with adsorption capacities (Krishna et al. 2023)

Adsorbent type	Heavy Metal Ion (HMI)	Adsorption capacity (mg/g)	References
MWCNTs-KOH@NiNPs	As, Cd, Pb	440.9, 415.8, 481	Egbosiuba et al. (2022)
$\text{MWCNTs}_{5\text{-}15\ \text{nm}}$	Pb	1306	Egbosiuba et al. (2021)
Polyethylene (PE)—Lignin (LG)—CNTs—Ni/Al_2O_3	Pb	146.08	Wang et al. (2021)
HNO_3 oxidized: MWCNTs 3 h	Cd	10.5	Šolić et al. (2021)
HNO_3 oxidized: MWCNTs 6 h	Pb	23.4	Šolić et al. (2021)
Chitosan-modified magnetic carbon nanotubes (CS/ MWCNTs/ Fe)	Cr	1.54	Aslam et al. (2021)
Carbon nanotube/ poly-amidoaminedendrimer composite (PAMAM/CNT)	As	432	Hayati et al. (2018)
Carboxylated multi-walled carbon nanotubes (MWCNTs-COOH)	Cd	212.76	Alimohammady et al. (2017)
Thiolated multi-walled carbon nanotubes (MWCNTs-SH)	Hg	204.64	Hadavifar et al. (2016)
Multi-walled carbon nanotube–hydroxyapatite composite (MWCNT–HAP)	Co	16.26	Liu et al. (2013)
Sulphur containing multi-walled carbon nanotubes (S-MWCNTs)	Hg	0.0728	Pillay et al. (2013)

1.4 Conclusions

A sustainable and efficient strategy for the remediation of heavy metals from the environment is the topmost priority due to the deadly impacts of these metals on health and the environment. Graphene, owing to unique attributes such as high surface-to-volume ratio, superior electron conductivity, great mechanical strength, flexibility for modifications and functionalization, has emerged as a super-material for environmental applications. As compared to conventional carbon-based adsorbents, graphene exhibits high efficiency in adsorptive removal of HMIS across the ecosystems. Besides, graphene derivatives such as GO and rGO, in addition to the large surface area, also possess numerous oxygen-derived surface functional groups, providing a large number of active sites for adsorption of HMS. The facile, time and cost-effective synthesis practices of graphene further add value for its environmental applications. The forthcoming chapters will cover the strategies involved in

Table 1.2 Represents graphene and its functionalized structures for the removal of different HMIS along with adsorption capacities (Ali et al. 2019)

Adsorbent type	Heavy Metal Ion (HMI)	Adsorption capacity (mg/g)	References
Graphene	Pb	22.42	Wang and Peng (2010)
Graphene (heat treated at 773 K)	Pb	35.21	Wang and Peng (2010)
Graphene (heat treated at 973 K)	Pb	35.46	Huang et al. (2011)
Graphene	Sb	10.919	Huang et al. (2011)
Graphene	Fe	299.3	Chang et al. (2013)
Graphene	Co	370	Chang et al. (2013)
CTAB modified graphene	Cr	21.57	Yan et al. (2012)
GO	Pb	367	Madadrang et al. (2012)
GO	Zn	30.1	Lee and Yang (2012)
GO	Cd	14.9	Lee and Yang (2012)
GO	Cu	117.5	Wu et al. (2013a, b, c)
GO aerogel	Cu	17.73	Mi et al. (2012)
EDTA modified GO	Pb	525	Madadrang et al. (2012)
FGO	Pb	842	Zhao et al. (2011a, b, c, d)
FGO	Cd	106.3	Zhao et al. (2011a, b, c, d)
FGO	Co	68.2	Zhao et al. (2011a, b, c, d)
Graphene/Fe	Cr	162	Jabeen et al. (2011)
Graphene/δ-MnO_2	Ni	46.55	Ren et al. (2011)
GO–TiO_2	Zn	88.9	Ren et al. (2011)
GO–TiO_2	Cd	72.8	Ren et al. (2011)
GO– TiO_2	Pb	65.6	Ren et al. (2011)

the adsorption of HMIS by pristine graphene and its derivatives, as well as functionalization techniques for graphene and applications of functionalized graphene in the HM remediation.

References

Abbas Q, Shinde PA, Abdelkareem MA, Alami AH, Mirzaeian M, Yadav A, Olabi AG (2022) Graphene synthesis techniques and environmental applications. Materials 15(21):7804. https://doi.org/10.3390/ma15217804

Abraham J, Jose B, Jose A, Thomas S (2020) Characterization of green nanoparticles from plants. In: Phytonanotechnology, Elsevier, pp 21–39. https://doi.org/10.1016/B978-0-12-822348-2.00002-4

Ahmad SZN, Wan Salleh WN, Ismail AF, Yusof N, Mohd Yusop MZ, Aziz F (2020) Adsorptive removal of heavy metal ions using graphene-based nanomaterials: Toxicity, roles of functional groups and mechanisms. Chemosphere 248:126008. https://doi.org/10.1016/j.chemosphere.2020.126008

Al Moustafa A-E, Mfoumou E, Roman D, Nerguizian V, Alazzam A, Stiharu I, Yasmeen A (2016) Impact of single-walled carbon nanotubes on the embryo: A brief review. Int J Nanomedicine:349. https://doi.org/10.2147/IJN.S96361

Alagappan PN, Heimann J, Morrow L, Andreoli E, Barron AR (2017) Easily regenerated readily deployable absorbent for heavy metal removal from contaminated water. Sci Rep 7(1):6682. https://doi.org/10.1038/s41598-017-06734-7

Al-Gaashani R, Najjar A, Zakaria Y, Mansour S, Atieh MA (2019) XPS and structural studies of high quality graphene oxide and reduced graphene oxide prepared by different chemical oxidation methods. Ceram Int 45(11):14439–14448. https://doi.org/10.1016/j.ceramint.2019.04.165

Ali H, Khan E (2018) What are heavy metals? Long-standing controversy over the scientific use of the term 'heavy metals'—proposal of a comprehensive definition. Toxicol Environ Chem 100(1):6–19. https://doi.org/10.1080/02772248.2017.1413652

Ali I, Basheer AA, Mbianda XY, Burakov A, Galunin E, Burakova I, Mkrtchyan E, Tkachev A, Grachev V (2019) Graphene based adsorbents for remediation of noxious pollutants from wastewater. Environ Int 127:160–180. https://doi.org/10.1016/j.envint.2019.03.029

Alimohammady M, Jahangiri M, Kiani F, Tahermansouri H (2017) A new modified MWCNTs with 3-aminopyrazole as a nanoadsorbent for Cd(II) removal from aqueous solutions. J Environ Chem Eng 5(4):3405–3417. https://doi.org/10.1016/j.jece.2017.06.045

Al-Mulla T, Qin Z, Buehler MJ (2015) Crumpling deformation regimes of monolayer graphene on substrate: A molecular mechanics study. J Phys: Condens Matter 27(34):345401. https://doi.org/10.1088/0953-8984/27/34/345401

Aragaw TA, Bogale FM, Aragaw BA (2021) Iron-based nanoparticles in wastewater treatment: A review on synthesis methods, applications, and removal mechanisms. J Saudi Chem Soc 25(8):101280. https://doi.org/10.1016/j.jscs.2021.101280

Aslam MMA, Den W, Kuo H-W (2021) Removal of hexavalent chromium by encapsulated chitosan-modified magnetic carbon nanotubes: Fixed-bed column study and modelling. J Water Process Eng 42:102143. https://doi.org/10.1016/j.jwpe.2021.102143

Beshkova M, Hultman L, Yakimova R (2016) Device applications of epitaxial graphene on silicon carbide. Vacuum 128:186–197. https://doi.org/10.1016/j.vacuum.2016.03.027

Bolotin KI, Sikes KJ, Jiang Z, Klima M, Fudenberg G, Hone J, Kim P, Stormer HL (2008) Ultrahigh electron mobility in suspended graphene. Solid State Commun 146(9–10):351–355. https://doi.org/10.1016/j.ssc.2008.02.024

Borji H, Ayoub GM, Bilbeisi R, Nassar N, Malaeb L (2020) How effective are nanomaterials for the removal of heavy metals from water and wastewater? Water Air Soil Pollut 231(7):330. https://doi.org/10.1007/s11270-020-04681-0

Brennen CE (2013) Cavitation and bubble dynamics. Cambridge University Press. https://doi.org/10.1017/CBO9781107338760

Briffa J, Sinagra E, Blundell R (2020) Heavy metal pollution in the environment and their toxicological effects on humans. Heliyon 6(9):e04691. https://doi.org/10.1016/j.heliyon.2020.e04691

Bruna T, Maldonado-Bravo F, Jara P, Caro N (2021) Silver nanoparticles and their antibacterial applications. Int J Mol Sci 22(13):7202. https://doi.org/10.3390/ijms22137202
Cai X, Jiang Z, Zhang X, Zhang X (2018) Effects of tip sonication parameters on liquid phase exfoliation of graphite into graphene nanoplatelets. Nanoscale Res Lett 13(1):241. https://doi.org/10.1186/s11671-018-2648-5
Casiraghi C, Hartschuh A, Lidorikis E, Qian H, Harutyunyan H, Gokus T, Novoselov KS, Ferrari AC (2007) Rayleigh imaging of graphene and graphene layers. Nano Lett 7(9):2711–2717. https://doi.org/10.1021/nl071168m
Chang C-F, Truong QD, Chen J-R (2013) RETRACTED: Graphene sheets synthesized by ionic-liquid-assisted electrolysis for application in water purification. Appl Surf Sci 264:329–334. https://doi.org/10.1016/j.apsusc.2012.10.022
Chen L, Yu H, Zhong J, Song L, Wu J, Su W (2017) Graphene field emitters: A review of fabrication, characterization and properties. Mater Sci Eng, B 220:44–58. https://doi.org/10.1016/j.mseb.2017.03.007
Choi W, Lahiri I, Seelaboyina R, Kang YS (2010) Synthesis of graphene and its applications: A review. Crit Rev Solid State Mater Sci 35(1):52–71. https://doi.org/10.1080/10408430903505036
Compton OC, Nguyen ST (2010) Graphene oxide, highly reduced graphene oxide, and graphene: versatile building blocks for carbon-based materials. Small 6(6):711–723. https://doi.org/10.1002/smll.200901934
De Marchi L, Pretti C, Gabriel B, Marques PAAP, Freitas R, Neto V (2018) An overview of graphene materials: Properties, applications and toxicity on aquatic environments. Sci Total Environ 631–632:1440–1456. https://doi.org/10.1016/j.scitotenv.2018.03.132
De Silva M, Cao G, Tam KC (2025) Nanomaterials for the removal and detection of heavy metals: A review. Environ Sci Nano 12(4):2154–2176. https://doi.org/10.1039/D4EN01041H
Diallo MS, Christie S, Swaminathan P, Johnson JH, Goddard WA (2005) Dendrimer enhanced ultrafiltration. 1. Recovery of Cu(II) from aqueous solutions using PAMAM dendrimers with ethylene diamine core and terminal NH_2 Groups. Environ Sci Technol 39(5):1366–1377. https://doi.org/10.1021/es048961r
Dreyer DR, Park S, Bielawski CW, Ruoff RS (2010) The chemistry of graphene oxide. Chem Soc Rev 39(1):228–240. https://doi.org/10.1039/B917103G
Edward K, Mamun K, Narayan S, Assaf M, Rohindra D, Rathnayake U (2023) State-of-the-art graphene synthesis methods and environmental concerns. Appl Environ Soil Sci 2023:1–23. https://doi.org/10.1155/2023/8475504
Edwards RS, Coleman KS (2013) Graphene synthesis: Relationship to applications. Nanoscale 5(1):38–51. https://doi.org/10.1039/C2NR32629A
Egbosiuba TC, Abdulkareem AS, Tijani JO, Ani JI, Krikstolaityte V, Srinivasan M, Veksha A, Lisak G (2021) Taguchi optimization design of diameter-controlled synthesis of multi walled carbon nanotubes for the adsorption of Pb(II) and Ni(II) from chemical industry wastewater. Chemosphere 266:128937. https://doi.org/10.1016/j.chemosphere.2020.128937
Egbosiuba TC, Egwunyenga MC, Tijani JO, Mustapha S, Abdulkareem AS, Kovo AS, Krikstolaityte V, Veksha A, Wagner M, Lisak G (2022) Activated multi-walled carbon nanotubes decorated with zero valent nickel nanoparticles for arsenic, cadmium and lead adsorption from wastewater in a batch and continuous flow modes. J Hazard Mater 423:126993. https://doi.org/10.1016/j.jhazmat.2021.126993
Exley C (2013) Human exposure to aluminium. Environ. Sci.: Processes Impacts, 15(10):1807–1816. https://doi.org/10.1039/C3EM00374D
Farghali AA, Abdel Tawab HA, Abdel Moaty SA, Khaled R (2017) Functionalization of acidified multi-walled carbon nanotubes for removal of heavy metals in aqueous solutions. J Nanostructure Chem 7(2):101–111. https://doi.org/10.1007/s40097-017-0227-4
Feisal NAS, Kamaludin NH, Ahmad MA, Tengku Ibrahim TNB (2024) A comprehensive review of nanomaterials for efficient heavy metal ions removal in water treatment. J Water Process Eng 64:105566. https://doi.org/10.1016/j.jwpe.2024.105566

Filipoiu DC, Bungau SG, Endres L, Negru PA, Bungau AF, Pasca B, Radu A-F, Tarce AG, Bogdan MA, Behl T, Nechifor AC, ul Hassan SS, Tit DM (2022) Characterization of the toxicological impact of heavy metals on human health in conjunction with modern analytical methods. Toxics 10(12):716. https://doi.org/10.3390/toxics10120716

Fitzgerald WF, Clarkson TW (1991) Mercury and monomethylmercury: Present and future concerns. Environ Health Perspect 96:159–166. https://doi.org/10.1289/ehp.9196159

Fulke AB, Ratanpal S, Sonker S (2024) Understanding heavy metal toxicity: Implications on human health, marine ecosystems and bioremediation strategies. Mar Pollut Bull 206:116707. https://doi.org/10.1016/j.marpolbul.2024.116707

Gadipelli S, Guo ZX (2015) Graphene-based materials: Synthesis and gas sorption, storage and separation. Prog Mater Sci 69:1–60. https://doi.org/10.1016/j.pmatsci.2014.10.004

Gaffney J, Marley N (2014) In-depth review of atmospheric mercury: sources, transformations, and potential sinks. Energy Emiss Control Technol:1. https://doi.org/10.2147/EECT.S37038

Gao Z, Wang S, Berry J, Zhang Q, Gebhardt J, Parkin WM, Avila J, Yi H, Chen C, Hurtado-Parra S, Drndić M, Rappe AM, Srolovitz DJ, Kikkawa JM, Luo Z, Asensio MC, Wang F, Johnson ATC (2020) Large-area epitaxial growth of curvature-stabilized ABC trilayer graphene. Nat Commun 11(1):546. https://doi.org/10.1038/s41467-019-14022-3

Garg B, Bisht T, Ling Y-C (2014) Graphene-based nanomaterials as heterogeneous acid catalysts: A comprehensive perspective. Molecules 19(9):14582–14614. https://doi.org/10.3390/molecules190914582

Geim AK (2009) Graphene: Status and prospects. Science (1979), 324(5934):1530–1534. https://doi.org/10.1126/science.1158877

Gensemer RW, Playle RC (1999) The bioavailability and toxicity of aluminum in aquatic environments. Crit Rev Environ Sci Technol 29(4):315–450. https://doi.org/10.1080/10643389991259245

Ghany NAA, Elsherif SA, Handal HT (2017) Revolution of graphene for different applications: State-of-the-art. Surf Interfaces 9:93–106. https://doi.org/10.1016/j.surfin.2017.08.004

Goldstein JI, Newbury DE, Michael JR, Ritchie NWM, Scott JHJ, Joy DC (2018) Scanning electron microscopy and x-ray microanalysis. Springer, New York. https://doi.org/10.1007/978-1-4939-6676-9

Griffiths PR (1983) Fourier transform infrared spectrometry. Science (1979) 222(4621):297–302. https://doi.org/10.1126/science.6623077

Guerra FD, Attia MF, Whitehead DC, Alexis F (2018) Nanotechnology for environmental remediation: materials and applications. Molecules 23(7):1760. https://doi.org/10.3390/molecules23071760

Hadavifar M, Bahramifar N, Younesi H, Rastakhiz M, Li Q, Yu J, Eftekhari E (2016) Removal of mercury(II) and cadmium(II) ions from synthetic wastewater by a newly synthesized amino and thiolated multi-walled carbon nanotubes. J Taiwan Inst Chem Eng 67:397–405. https://doi.org/10.1016/j.jtice.2016.08.029

Hayati B, Maleki A, Najafi F, Gharibi F, McKay G, Gupta VK, Harikaranahalli Puttaiah S, Marzban N (2018) Heavy metal adsorption using PAMAM/CNT nanocomposite from aqueous solution in batch and continuous fixed bed systems. Chem Eng J 346:258–270. https://doi.org/10.1016/j.cej.2018.03.172

Hayati B, Maleki A, Najafi F, Daraei H, Gharibi F, McKay G (2017) Super high removal capacities of heavy metals (Pb 2+ and Cu 2+) using CNT dendrimer. J Hazard Mater 336:146–157. https://doi.org/10.1016/j.jhazmat.2017.02.059

He Y, Fang H, Pan X, Zhu B, Chen J, Wang J, Zhang R, Chen L, Qi X, Zhang H (2023) Cadmium exposure in aquatic products and health risk classification assessment in residents of Zhejiang, China. Foods 12(16):3094. https://doi.org/10.3390/foods12163094

Hontoria-Lucas C, López-Peinado AJ, de López-González JD, Rojas-Cervantes ML, Martín-Aranda RM (1995) Study of oxygen-containing groups in a series of graphite oxides: Physical and chemical characterization. Carbon NY, 33(11):1585–1592. https://doi.org/10.1016/0008-6223(95)00120-3

Huang B, Wang Q, Li Y, Zhang M, Wei X (2015) Preparation and characterisation of graphene. Mater Res Innov 19(sup9):S9–344–S9–350. https://doi.org/10.1179/1432891715Z.0000000002010
Huang HY, Yang RT, Chinn D, Munson CL (2003) Amine-grafted MCM-48 and silica xerogel as superior sorbents for acidic gas removal from natural gas. Ind Eng Chem Res 42(12):2427–2433. https://doi.org/10.1021/ie020440u
Huang Z-H, Zheng X, Lv W, Wang M, Yang Q-H, Kang F (2011) Adsorption of lead(II) ions from aqueous solution on low-temperature exfoliated graphene nanosheets. Langmuir 27(12):7558–7562. https://doi.org/10.1021/la200606r
Igbokwe IO, Igwenagu E, Igbokwe NA (2019) Aluminium toxicosis: a review of toxic actions and effects. Interdiscip Toxicol 12(2):45–70. https://doi.org/10.2478/intox-2019-0007
Ihsanullah, Abbas A, Al-Amer AM, Laoui T, Al-Marri MJ, Nasser MS, Khraisheh M, Atieh MA (2016) Heavy metal removal from aqueous solution by advanced carbon nanotubes: Critical review of adsorption applications. Sep Purif Technol 157:141–161. https://doi.org/10.1016/j.seppur.2015.11.039
Jabeen H, Chandra V, Jung S, Lee JW, Kim KS, Bin Kim S (2011) Enhanced Cr(vi) removal using iron nanoparticle decorated graphene. Nanoscale 3(9):3583. https://doi.org/10.1039/c1nr10549c
Johra FT, Lee J-W, Jung W-G (2014) Facile and safe graphene preparation on solution based platform. J Ind Eng Chem 20(5):2883–2887. https://doi.org/10.1016/j.jiec.2013.11.022
Jomova K, Alomar SY, Nepovimova E, Kuca K, Valko M (2025) Heavy metals: Toxicity and human health effects. Arch Toxicol 99(1):153–209. https://doi.org/10.1007/s00204-024-03903-2
Jones FT (2007) A broad view of arsenic. Poult Sci 86(1):2–14. https://doi.org/10.1093/ps/86.1.2
Kaniyoor A, Ramaprabhu S (2012) A Raman spectroscopic investigation of graphite oxide derived graphene. AIP Adv 2(3). https://doi.org/10.1063/1.4756995
Karthik V, Selvakumar P, Senthil Kumar P, Vo D-VN, Gokulakrishnan M, Keerthana P, Tamil Elakkiya V, Rajeswari R (2021) Graphene-based materials for environmental applications: A review. Environ Chem Lett 19(5):3631–3644. https://doi.org/10.1007/s10311-021-01262-3
Karu AE, Beer M (1966) Pyrolytic formation of highly crystalline graphite films. J Appl Phys 37(5):2179–2181. https://doi.org/10.1063/1.1708759
Kim KS, Zhao Y, Jang H, Lee SY, Kim JM, Kim KS, Ahn J-H, Kim P, Choi J-Y, Hong BH (2009) Large-scale pattern growth of graphene films for stretchable transparent electrodes. Nature 457(7230):706–710. https://doi.org/10.1038/nature07719
Knieke C, Berger A, Peukert W (2010) Graphene production with stirred media mills. MRS Proceedings 1259:1259–S12–01. https://doi.org/10.1557/PROC-1259-S12-01
Konwar A, Kalita S, Kotoky J, Chowdhury D (2016) Chitosan–iron oxide coated graphene oxide nanocomposite hydrogel: A robust and soft antimicrobial biofilm. ACS Appl Mater Interfaces 8(32):20625–20634. https://doi.org/10.1021/acsami.6b07510
Krishna RH, Chandraprabha MN, Samrat K, Krishna Murthy TP, Manjunatha C, Kumar SG (2023) Carbon nanotubes and graphene-based materials for adsorptive removal of metal ions—A review on surface functionalization and related adsorption mechanism. Appl Surf Sci Adv 16:100431. https://doi.org/10.1016/j.apsadv.2023.100431
Kumar N, Salehiyan R, Chauke V, Joseph Botlhoko O, Setshedi K, Scriba M, Masukume M, Sinha Ray S (2021) Top-down synthesis of graphene: A comprehensive review. FlatChem 27:100224. https://doi.org/10.1016/j.flatc.2021.100224
Lee C, Wei X, Kysar JW, Hone J (2008) Measurement of the elastic properties and intrinsic strength of monolayer graphene. Science (1979), 321(5887):385–388. https://doi.org/10.1126/science.1157996
Lee Y-C, Yang J-W (2012) Self-assembled flower-like TiO_2 on exfoliated graphite oxide for heavy metal removal. J Ind Eng Chem 18(3):1178–1185. https://doi.org/10.1016/j.jiec.2012.01.005
Li J, Pandey GP (2015) Advanced physical chemistry of carbon nanotubes. Annu Rev Phys Chem 66(1):331–356. https://doi.org/10.1146/annurev-physchem-040214-121535

Li X, Cai W, Jung IH, An JH, Yang D, Velamakanni A, Piner R, Colombo L, Ruoff RS (2009) Synthesis, characterization, and properties of large-area graphene films. ECS Trans 19(5):41–52. https://doi.org/10.1149/1.3119526

Lin Y-M, Dimitrakopoulos C, Jenkins KA, Farmer DB, Chiu H-Y, Grill A, Avouris P (2010) 100-GHz transistors from wafer-scale epitaxial graphene. Science (1979), 327(5966):662–662. https://doi.org/10.1126/science.1184289

Lin L, Peng H, Liu Z (2019) Synthesis challenges for graphene industry. Nat Mater 18(6):520–524. https://doi.org/10.1038/s41563-019-0341-4

Liu F, Wang C, Sui X, Riaz MA, Xu M, Wei L, Chen Y (2019) Synthesis of graphene materials by electrochemical exfoliation: Recent progress and future potential. Carbon Energy 1(2):173–199. https://doi.org/10.1002/cey2.14

Liu Z, Chen L, Zhang Z, Li Y, Dong Y, Sun Y (2013) Synthesis of multi-walled carbon nanotube–hydroxyapatite composites and its application in the sorption of Co(II) from aqueous solutions. J Mol Liq 179:46–53. https://doi.org/10.1016/j.molliq.2012.12.011

Madadrang CJ, Kim HY, Gao G, Wang N, Zhu J, Feng H, Gorring M, Kasner ML, Hou S (2012) Adsorption behavior of EDTA-graphene oxide for Pb (II) removal. ACS Appl Mater Interfaces 4(3):1186–1193. https://doi.org/10.1021/am201645g

Mahajan P, Khanna V (2025) Nanoparticle-mediated regulation of chromium toxicity in plants: Unveiling the mechanism at cellular level, pp 269–290. https://doi.org/10.1007/978-3-031-77327-3_14

Mahajan P, Batish DR, Singh HP, Kohli RK (2016) β-Pinene partially ameliorates Cr(VI)-inhibited growth and biochemical changes in emerging seedlings. Plant Growth Regul 79(2):243–249. https://doi.org/10.1007/s10725-015-0129-8

Mahajan P, Sharma P, Singh HP, Rathee S, Sharma M, Batish DR, Kohli RK (2021) Amelioration potential of β-pinene on Cr(VI)-induced toxicity on morphology, physiology and ultrastructure of maize. Environ Sci Pollut Res 28(44):62431–62443. https://doi.org/10.1007/s11356-021-15018-7

Mahajan P, Singh HP, Kaur S, Batish DR, Kohli RK (2019) β-Pinene moderates Cr(VI) phytotoxicity by quenching reactive oxygen species and altering antioxidant machinery in maize. Environ Sci Pollut Res 26(1):456–463. https://doi.org/10.1007/s11356-018-3562-1

Mahdavi S, Jalali M, Afkhami A (2013) Heavy metals removal from aqueous solutions using TiO_2, MgO, and Al_2O_3 nanoparticles. Chem Eng Commun 200(3):448–470. https://doi.org/10.1080/00986445.2012.686939

Mauter MS, Elimelech M (2008) Environmental applications of carbon-based nanomaterials. Environ Sci Technol 42(16):5843–5859. https://doi.org/10.1021/es8006904

Mbayachi VB, Ndayiragije E, Sammani T, Taj S, Mbuta ER, ullah Khan A (2021) Graphene synthesis, characterization and its applications: A review. Results Chem 3:100163. https://doi.org/10.1016/j.rechem.2021.100163

Mi X, Huang G, Xie W, Wang W, Liu Y, Gao J (2012) Preparation of graphene oxide aerogel and its adsorption for Cu2+ ions. Carbon N Y 50(13):4856–4864. https://doi.org/10.1016/j.carbon.2012.06.013

Mishra N, Boeckl J, Motta N, Iacopi F (2016) Graphene growth on silicon carbide: A review. Phys Status Solidi (A) 213(9):2277–2289. https://doi.org/10.1002/pssa.201600091

Moreno-Fernández G, Boulanger N, Nordenström A, Iakunkov A, Talyzin A, Carriazo D, Mysyk R (2021) Ball-milling-enhanced capacitive charge storage of activated graphene in aqueous, organic and ionic liquid electrolytes. Electrochim Acta 370:137738. https://doi.org/10.1016/j.electacta.2021.137738

Mubarak N, Thobashinni M, Abdullah E, Sahu J (2016) Comparative kinetic study of removal of Pb2+ ions and Cr3+ ions from waste water using carbon nanotubes produced using microwave heating. C (Basel) 2(1):7. https://doi.org/10.3390/c2010007

Mudzielwana R, Gitari MW, Ndungu P (2019) Uptake of As(V) from groundwater using Fe-Mn oxides modified kaolin clay: physicochemical characterization and adsorption data modeling. Water (Basel) 11(6):1245. https://doi.org/10.3390/w11061245

Nayl AA, Abd-Elhamid AI, Aly AA, Bräse S (2022) Recent progress in the applications of silica-based nanoparticles. RSC Adv 12(22):13706–13726. https://doi.org/10.1039/D2RA01587K
Novoselov KS, Fal′ko VI, Colombo L, Gellert PR, Schwab MG, Kim K (2012) A roadmap for graphene. Nature 490(7419):192–200. https://doi.org/10.1038/nature11458
Novoselov KS, Jiang D, Schedin F, Booth TJ, Khotkevich VV, Morozov SV, Geim AK (2005) Two-dimensional atomic crystals. Proc Natl Acad Sci 102(30):10451–10453. https://doi.org/10.1073/pnas.0502848102
Nuvoli D, Valentini L, Alzari V, Scognamillo S, Bon SB, Piccinini M, Illescas J, Mariani A (2011) High concentration few-layer graphene sheets obtained by liquid phase exfoliation of graphite in ionic liquid. J Mater Chem 21(10):3428–3431. https://doi.org/10.1039/C0JM02461A
Papageorgiou DG, Kinloch IA, Young RJ (2017) Mechanical properties of graphene and graphene-based nanocomposites. Prog Mater Sci 90:75–127. https://doi.org/10.1016/j.pmatsci.2017.07.004
Park S, Ruoff RS (2009) Chemical methods for the production of graphenes. Nat Nanotechnol 4(4):217–224. https://doi.org/10.1038/nnano.2009.58
Paton KR, Varrla E, Backes C, Smith RJ, Khan U, O'Neill A, Boland C, Lotya M, Istrate OM, King P, Higgins T, Barwich S, May P, Puczkarski P, Ahmed I, Moebius M, Pettersson H, Long E, Coelho J, O'Brien SE, McGuire EK, Sanchez BM, Duesberg GS, McEvoy N, Pennycook TJ, Downing C, Crossley A, Nicolosi V, Coleman JN (2014) Scalable production of large quantities of defect-free few-layer graphene by shear exfoliation in liquids. Nat Mater 13(6):624–630. https://doi.org/10.1038/nmat3944
Peigney A, Laurent C, Flahaut E, Bacsa RR, Rousset A (2001) Specific surface area of carbon nanotubes and bundles of carbon nanotubes. Carbon N Y 39(4):507–514. https://doi.org/10.1016/S0008-6223(00)00155-X
Phaenark C, Phankamolsil Y, Sawangproh W (2024) Ecological and health implications of heavy metal bioaccumulation in Thai Fauna: A systematic review. Ecotoxicol Environ Saf 285:117086. https://doi.org/10.1016/j.ecoenv.2024.117086
Pillay K, Cukrowska EM, Coville NJ (2013) Improved uptake of mercury by sulphur-containing carbon nanotubes. Microchem J 108:124–130. https://doi.org/10.1016/j.microc.2012.10.014
Prekodravac JR, Kepić DP, Colmenares JC, Giannakoudakis DA, Jovanović SP (2021) A comprehensive review on selected graphene synthesis methods: From electrochemical exfoliation through rapid thermal annealing towards biomass pyrolysis. J Mater Chem C Mater 9(21):6722–6748. https://doi.org/10.1039/D1TC01316E
Qin Z, Taylor M, Hwang M, Bertoldi K, Buehler MJ (2014) Effect of wrinkles on the surface area of graphene: Toward the design of nanoelectronics. Nano Lett 14(11):6520–6525. https://doi.org/10.1021/nl503097u
Recycling used lead-acid batteries: health considerations
Rabchinskii MK, Ryzhkov SA, Kirilenko DA, Ulin NV, Baidakova MV, Shnitov VV, Pavlov SI, Chumakov RG, Stolyarova DYU, Besedina NA, Shvidchenko AV, Potorochin DV, Roth F, Smirnov DA, Gudkov MV, Brzhezinskaya M, Lebedev OI, Melnikov VP, Brunkov PN (2020) From graphene oxide towards aminated graphene: facile synthesis, its structure and electronic properties. Sci Rep 10(1):6902. https://doi.org/10.1038/s41598-020-63935-3
Ren X, Chen C, Nagatsu M, Wang X (2011) Carbon nanotubes as adsorbents in environmental pollution management: A review. Chem Eng J 170(2–3):395–410. https://doi.org/10.1016/j.cej.2010.08.045
Ren Y, Yan N, Wen Q, Fan Z, Wei T, Zhang M, Ma J (2011) Graphene/δ-MnO_2 composite as adsorbent for the removal of nickel ions from wastewater. Chem Eng J 175:1–7. https://doi.org/10.1016/j.cej.2010.08.010
Robinson AL (1986) Congress Critical of foot-dragging on critical materials. Science (1979) 234(4772):20–21. https://doi.org/10.1126/science.234.4772.20.b
Sahu TK, Ranjan P, Kumar P (2021) Chemical exfoliation synthesis of boron nitride and molybdenum disulfide 2D sheets via modified Hummers' method. Emergent Mater 4(3):645–654. https://doi.org/10.1007/s42247-021-00170-0

Sanchez VC, Jachak A, Hurt RH, Kane AB (2012) Biological interactions of graphene-family nanomaterials: An interdisciplinary review. Chem Res Toxicol 25(1):15–34. https://doi.org/10.1021/tx200339h

Seekaew Y, Lokavee S, Phokharatkul D, Wisitsoraat A, Kerdcharoen T, Wongchoosuk C (2014) Low-cost and flexible printed graphene–PEDOT: PSS gas sensor for ammonia detection. Org Electron 15(11):2971–2981. https://doi.org/10.1016/j.orgel.2014.08.044

Seid Ali Y, Shaw I, Liu Y, Chen C (2024) Perspective chapter: Advanced nanotechnology approach for heavy metal toxicity—analysis, treatment, and removal. In: Heavy metals in the environment—contamination, risk, and remediation, IntechOpen. https://doi.org/10.5772/intechopen.1006734

Shah J, Lopez-Mercado J, Carreon MG, Lopez-Miranda A, Carreon ML (2018) Plasma synthesis of graphene from mango peel. ACS Omega 3(1):455–463. https://doi.org/10.1021/acsomega.7b01825

Shang W, Cai T, Zhang Y, Liu D, Liu S (2018) Facile one pot pyrolysis synthesis of carbon quantum dots and graphene oxide nanomaterials: All carbon hybrids as eco-environmental lubricants for low friction and remarkable wear-resistance. Tribol Int 118:373–380. https://doi.org/10.1016/j.triboint.2017.09.029

Sheela T, Nayaka YA, Viswanatha R, Basavanna S, Venkatesha TG (2012) Kinetics and thermodynamics studies on the adsorption of Zn(II), Cd(II) and Hg(II) from aqueous solution using zinc oxide nanoparticles. Powder Technol 217:163–170. https://doi.org/10.1016/j.powtec.2011.10.023

Singh HP, Mahajan P, Kaur S, Batish DR, Kohli RK (2013) Chromium toxicity and tolerance in plants. Environ Chem Lett 11(3):229–254. https://doi.org/10.1007/s10311-013-0407-5

Šolić M, Maletić S, Isakovski MK, Nikić J, Watson M, Kónya Z, Rončević S (2021) Removing low levels of Cd(II) and Pb(II) by adsorption on two types of oxidized multiwalled carbon nanotubes. J Environ Chem Eng 9(4):105402. https://doi.org/10.1016/j.jece.2021.105402

Sreeprasad TS, Berry V (2013) How do the electrical properties of graphene change with its functionalization? Small 9(3):341–350. https://doi.org/10.1002/smll.201202196

Su C-Y, Lu A-Y, Xu Y, Chen F-R, Khlobystov AN, Li L-J (2011) High-quality thin graphene films from fast electrochemical exfoliation. ACS Nano 5(3):2332–2339. https://doi.org/10.1021/nn200025p

Sumdani MG, Islam MR, Yahaya ANA, Safie SI (2021) Recent advances of the graphite exfoliation processes and structural modification of graphene: A review. J Nanopart Res 23(11):253. https://doi.org/10.1007/s11051-021-05371-6

Sun L, Yuan G, Gao L, Yang J, Chhowalla M, Gharahcheshmeh MH, Gleason KK, Choi YS, Hong BH, Liu Z (2021) Chemical vapour deposition. Nat Rev Methods Prim 1(1):5. https://doi.org/10.1038/s43586-020-00005-y

Suslick KS, Eddingsaas NC, Flannigan DJ, Hopkins SD, Xu H (2011) Extreme conditions during multibubble cavitation: Sonoluminescence as a spectroscopic probe. Ultrason Sonochem 18(4):842–846. https://doi.org/10.1016/j.ultsonch.2010.12.012

Syam Sundar L, Amin Mir M, Waqar Ashraf M, Djavanroodi F (2023) Synthesis and characterization of graphene and its composites for Lithium-Ion battery applications: A comprehensive review. Alex Eng J 78:224–245. https://doi.org/10.1016/j.aej.2023.07.044

The reflection of X-rays by crystals. In: Proceedings of the royal society of London. Series A, containing papers of a mathematical and physical character, vol 88, no 605, pp 428–438. https://doi.org/10.1098/rspa.1913.0040

Torres-Mendieta R, Ventura-Espinosa D, Sabater S, Lancis J, Mínguez-Vega G, Mata JA (2016) In situ decoration of graphene sheets with gold nanoparticles synthetized by pulsed laser ablation in liquids. Sci Rep 6(1):30478. https://doi.org/10.1038/srep30478

Trivedi S, Lobo K, Ramakrishna Matte HSS (2019) Synthesis, properties, and applications of graphene. In: Fundamentals and sensing applications of 2d materials, Elsevier, pp 25–90. https://doi.org/10.1016/B978-0-08-102577-2.00003-8

Valko M, Morris H, Cronin M (2005) Metals, toxicity and oxidative stress. Curr Med Chem 12(10):1161–1208. https://doi.org/10.2174/0929867053764635

Van Noorden R (2012) Production: Beyond sticky tape. Nature 483(7389):S32–S33. https://doi.org/10.1038/483S32a

Wan X, Huang Y, Chen Y (2012) Focusing on energy and optoelectronic applications: A journey for graphene and graphene oxide at large scale. Acc Chem Res 45(4):598–607. https://doi.org/10.1021/ar200229q

Wang S, Peng Y (2010) Natural zeolites as effective adsorbents in water and wastewater treatment. Chem Eng J 156(1):11–24. https://doi.org/10.1016/j.cej.2009.10.029

Wang X-Y, Narita A, Müllen K (2017) Precision synthesis versus bulk-scale fabrication of graphenes. Nat Rev Chem 2(1):0100. https://doi.org/10.1038/s41570-017-0100

Wang Z, Qin K, Wang Z, Shen D, Wu C (2021) Carbon nanotubes/Al_2O_3 composite derived from catalytic reforming of the pyrolysis volatiles of the mixture of polyethylene and lignin for highly-efficient removal of Pb(<scp>ii</scp>). RSC Adv 11(60):37851–37865. https://doi.org/10.1039/D1RA06762A

Wani AL, Ara A, Usmani JA (2015) Lead toxicity: A review. Interdiscip Toxicol 8(2):55–64. https://doi.org/10.1515/intox-2015-0009

Wu W, Yang Y, Zhou H, Ye T, Huang Z, Liu R, Kuang Y (2013a) Highly efficient removal of Cu(II) from aqueous solution by using graphene oxide. Water Air Soil Pollut 224(1):1372. https://doi.org/10.1007/s11270-012-1372-5

Wu Y, Luo H, Wang H, Wang C, Zhang J, Zhang Z (2013b) Adsorption of hexavalent chromium from aqueous solutions by graphene modified with cetyltrimethylammonium bromide. J Colloid Interface Sci 394:183–191. https://doi.org/10.1016/j.jcis.2012.11.049

Wu Q, Feng C, Wang C, Wang Z (2013c) A facile one-pot solvothermal method to produce superparamagnetic graphene–Fe3O4 nanocomposite and its application in the removal of dye from aqueous solution. Colloids Surf B Biointerfaces 101:210–214. https://doi.org/10.1016/j.colsurfb.2012.05.036

Xu Y, Cao H, Xue Y, Li B, Cai W (2018) Liquid-phase exfoliation of graphene: An overview on exfoliation media, techniques, and challenges. Nanomaterials 8(11):942. https://doi.org/10.3390/nano8110942

Yan L, Zheng YB, Zhao F, Li S, Gao X, Xu B, Weiss PS, Zhao Y (2012) Chemistry and physics of a single atomic layer: strategies and challenges for functionalization of graphene and graphene-based materials. Chem Soc Rev 41(1):97–114. https://doi.org/10.1039/C1CS15193B

Yang S, Lohe MR, Müllen K, Feng X (2016) New-generation graphene from electrochemical approaches: Production and applications. Adv Mater 28(29):6213–6221. https://doi.org/10.1002/adma.201505326

Yang W, Ratinac KR, Ringer SP, Thordarson P, Gooding JJ, Braet F (2010) Carbon nanomaterials in biosensors: should you use nanotubes or graphene? Angew Chem Int Ed 49(12):2114–2138. https://doi.org/10.1002/anie.200903463

Yang X, Shen Z, Zhang B, Yang J, Hong W-X, Zhuang Z, Liu J (2013) Silica nanoparticles capture atmospheric lead: Implications in the treatment of environmental heavy metal pollution. Chemosphere 90(2):653–656. https://doi.org/10.1016/j.chemosphere.2012.09.033

Yoon K-Y, Dong G (2020) Liquid-phase bottom-up synthesis of graphene nanoribbons. Mater Chem Front 4(1):29–45. https://doi.org/10.1039/C9QM00519F

Yousefi N, Lu X, Elimelech M, Tufenkji N (2019) Environmental performance of graphene-based 3D macrostructures. Nat Nanotechnol 14(2):107–119. https://doi.org/10.1038/s41565-018-0325-6

Zhang H-B, Zheng W-G, Yan Q, Yang Y, Wang J-W, Lu Z-H, Ji G-Y, Yu Z-Z (2010) Electrically conductive polyethylene terephthalate/graphene nanocomposites prepared by melt compounding. Polymer (Guildf) 51(5):1191–1196. https://doi.org/10.1016/j.polymer.2010.01.027

Zhang L, Li X, Huang Y, Ma Y, Wan X, Chen Y (2010) Controlled synthesis of few-layered graphene sheets on a large scale using chemical exfoliation. Carbon NY 48(8):2367–2371. https://doi.org/10.1016/j.carbon.2010.02.035

Zhang Y, Tan Y-W, Stormer HL, Kim P (2005) Experimental observation of the quantum Hall effect and Berry's phase in graphene. Nature 438(7065):201–204. https://doi.org/10.1038/nature04235

Zhao G, Jiang L, He Y, Li J, Dong H, Wang X, Hu W (2011c) Sulfonated graphene for persistent aromatic pollutant management. Adv Mater 23(34):3959–3963. https://doi.org/10.1002/adma.201101007

Zhao G, Li J, Ren X, Chen C, Wang X (2011a) Few-layered graphene oxide nanosheets as superior sorbents for heavy metal ion pollution management. Environ Sci Technol 45(24):10454–10462. https://doi.org/10.1021/es203439v

Zhao G, Li J, Wang X (2011d) Kinetic and thermodynamic study of 1-naphthol adsorption from aqueous solution to sulfonated graphene nanosheets. Chem Eng J 173(1):185–190. https://doi.org/10.1016/j.cej.2011.07.072

Zhao G, Ren X, Gao X, Tan X, Li J, Chen C, Huang Y, Wang X (2011c) Removal of Pb(ii) ions from aqueous solutions on few-layered graphene oxide nanosheets. Dalton Trans 40(41):10945. https://doi.org/10.1039/c1dt11005e

Zhao G, Wen T, Chen C, Wang X (2012) Synthesis of graphene-based nanomaterials and their application in energy-related and environmental-related areas. RSC Adv 2(25):9286. https://doi.org/10.1039/c2ra20990j

Chapter 2
Graphene-Based Adsorption Strategies for Heavy Metal Removal

2.1 Adsorption Mechanisms and Fundamentals

The process of adsorption refers to a bulk exchange mechanism that involves the accumulation of substances at the interface of two substrates/phases, viz., liquid–solid, liquid–liquid, and liquid–gas. The materials which allow the adherence of the components/substances are termed as adsorbents, and the components which get aggregated are known as adsorbates (Nayak et al. 2024). Adsorption phenomenon is generally segregated into physical and chemical adsorption, about the adsorption of gaseous molecules onto the solid substrate. The physical adsorption is governed by the weak interaction of gas molecules with the solid surface, similar to the forces which bind the molecules together in the liquid phase of the gas. Hence, their heats of condensation are comparable to the heats of adsorption, and both are relatively low. In other words, the gases which can be easily adsorbed can also be easily liquefied. These weak interactions are termed van der Waals forces, which are involved in the process of physical adsorption. Moreover, by decreasing the pressure, the process of physical adsorption can be reversed at the same temperature. Another adsorption process is chemical adsorption, which is governed by the chemical interaction forces operating between the adsorption surface and surrounding gaseous or solute molecules immersed in a liquid phase, alternatively termed as chemisorption. The process of chemical adsorption is not very prevalent, however, it is accompanied by an immense (nearly 10 times or greater) heat of adsorption. This process occurs more successfully at relevant temperatures, and for a reversible reaction accelerated temperature is required. Besides, the solute particles which tend to diminish surface tension will be aggregated at an interface between a solution and a solid substrate, exhibiting affinity to get adsorbed onto the solid surface. However, the solutes which aggravate the surface tension show a marked reduction in their concentration at the interface, termed as negative adsorption. For instance, most solutes possess a tendency to reduce the high surface tension of the water, thus taken easily by the adsorbent, however, alcohol possess less surface tension in comparison to water

P. Mahajan and V. Khanna, *Graphene in the Environment*, Environmental Science and Engineering, https://doi.org/10.1007/978-3-031-96708-5_2

which hinders the tendency of solutes to further lower it, hence it is difficult to adsorb the solutes from the alcoholic solution. This phenomenon of thermodynamics has been termed as "Gibbs adsorption equation". Besides, biosorption refers to the process of adsorption occurring on biological systems. This is highly significant in the removal and recovery of metals. The benefits conferred by biosorption include cost-effective, eco-friendly, and easily recoverable adsorbents such as industrial and agricultural waste, algae, fungi, bacteria, etc. (Raji et al. 2023).

2.1.1 Heavy Metal (HM) Ions Adsorption Mechanisms

Various conventional techniques have been applied for the remediation of Heavy metal (HM) ions, such as particle trade, layer detachment, filtration, ultrafiltration, electrodialysis, sedimentation, photocatalysis, flocculation, and adsorption (Rajendran et al. 2022). Most of these techniques have drawbacks linked to their cost, operational complexity, and high energy consumption. However, the adsorption mechanism has gained a lot of popularity due to its low operational cost, ease of conducting, high removal efficiency, simple set-up, and effortless adsorption of HM ion (Yang et al. 2019).

Figure 2.1 represents the general mechanism involved in the adsorption of HM ions from the wastewater.

The major mechanisms in the adsorption of HM ions on a particular substrate such as carbon-based substrate involve precipitation, electrostatic interaction, physical adsorption, surface complexation, and ion exchange (Fig. 2.2).

Precipitation: This phenomenon generates solid products in the process of HM ions adsorption, conducted either in the aqueous phase or on the surface (Inyang et al. 2012; Wang et al. 2018). For instance, biochar as an adsorbent for the removal of lead (Pb) ions produces minerals viz., cerussite ($Pbco_3$) and hydrocerussite [$Pb_3(CO_3)_2$] from them (Inyang et al. 2012, 2011). The precipitation exhibits much higher kinetics as compared to other techniques. Though precipitation is one of the key mechanisms

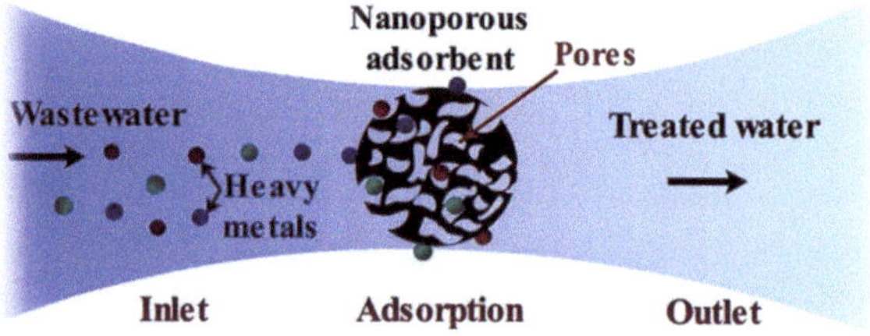

Fig. 2.1 Schematic representation of the heavy metal ion adsorption process for water treatment. Metal ions from wastewater adhere to the surface of nanoporous adsorbents, which possess a high surface area due to their porosity. The process can exhibit selectivity towards specific metal ions. Regeneration of the adsorbent is achieved using a suitable desorbing agent (Qasem et al. 2021)

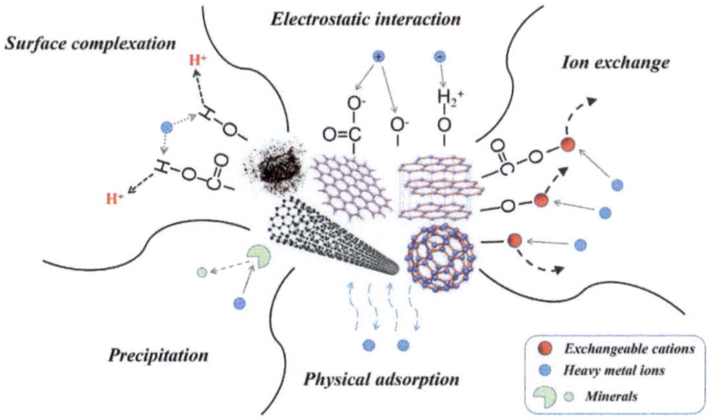

Fig. 2.2 illustrative view of adsorption techniques of HMs onto carbon adsorbents (Yang et al. 2019)

for HM remediation, however, it works in synchronisation with other adsorption techniques such as surface complexation, ion exchange, and electrostatic interaction (Inyang et al. 2016).

Electrostatic interaction: The underlying mechanism in this process is an association between positively charged HM ions and negatively charged carbon adsorbent, particularly in the presence of surface functional groups (Fu et al. 2016; Lv et al. 2018). Electrostatic interaction only has a little or secondary role in the removal of HM ions, as it is a relatively weak phenomenon (Cui et al. 2015). The adsorption of HM ions onto the carbon surface via this process completely depends upon the ph of the media and the point of zero charge of the adsorbing surface (Yang et al. 2019). Furthermore, surface group ionization supports a charged interface between the media and the carbon adsorbent. For instance, the interplay between ion exchange, surface complexation, and electrostatic interaction confers high adsorption efficiency to manganese oxide (Mno_2) biochar-based adsorbent for the removal of Cd and Pb ions (Liang et al. 2017).

Physical adsorption: This type of adsorption process does not involve any kind of bond formation, it is based upon the surface area of the adsorbent as well as pore size distribution. It is simply based upon HM ions' movement via diffusion into the pores present on the carbon adsorbent and then their deposition on their surface without the formation of any bonds. The enhanced micropores elevate the surface area, which assists the physical adsorption process; however, increased mesoporous supports HM ions diffusion, accelerating adsorption kinetics (Fomkin et al. 2017; Silva Buarque et al. 2015). Though HM ions removal through physical adsorption is a common practice, it is certainly not the predominant process for HM ions remediation, as it is weak. Although, few studies as one conducted by Nayak and co-workers, reported the importance of this process in controlling the adsorption of Ni and Cd from the aqueous media by two types of activated carbons in combination with the pore diffusion technique (Nayak et al. 2017).

Surface complexation: This involves the formation of multiatom structures or complexes via exclusive interaction with metal-functional groups (Peng et al. 2017) which facilitates HM ions adsorption onto carbon substrates (Liu et al. 2017). In this regard, it is observed that the biochar fabricated at low temperatures possess phenolic, lactonic, and carboxyl functional group which are able to adsorb HM ions with great efficiency (Mohan et al. 2007).

Ion exchange: It is one of the key mechanisms for removal of HM ions by carbon adsorbents, which involves exchange of ions between protons present in the oxygen-containing functional groups, viz., hydroxyl and carboxyl groups and the HM ions (El Kady etal. 2016). The robustness of the ion exchange technique is highly governed by the surface functional groups of the adsorbents and the ion particle size of the metal adsorbate (Dong et al. 2018). Besides, ph of the solution is another important factor which impacts the ion exchange process. A large number of protons are available at acidic ph, which saturates the metal binding sites. Moreover, releasing the proton from the adsorbent surface, which adsorbs the metal ions, changes the solution ph (Yang et al. 2019).

2.2 Graphene and Graphene Derivatives and Their Adsorption Characteristics

2.2.1 Graphene and Graphene Derivatives

As already discussed in the introduction part, graphene exhibits unique mechanical, electrical, and thermal properties. It is further classified into different categories such as two-dimensional (2D) graphene, which includes graphene oxide (GO) and reduced graphene oxide (rGO), three-dimensional (3d) graphene, viz., graphene hydrogels and graphene aerogels, graphene quantum dots (GQDS), and mesoporous graphene. These graphene-derived structures have gained a lot of attention in a short period due to their outstanding ability in environmental remediation and various other applications (Pan et al. 2020).

Graphene: As already described, graphene is represented by an atomic thick 2D layer of sp^2-bonded carbon atoms (Novoselov et al. 1979). It is gaining immense importance worldwide due to its magnificent electrical, optical, mechanical, and chemical attributes, viz., elevated transparency, effortless tempering, immense strength, and strong electrical conductivity (Novoselov et al. 2012). The peculiar properties of graphene enhance its applications in the field of energy storage and electronics. Besides, it is playing a significant role in environmental remediation, facilitating the adsorption of organic and inorganic pollutants. The adsorption excellence of graphene and its derivatives is governed by their hydrophobic attributes and outstanding bonding existing in their 2D aromatic structure, favouring adsorption of nagging aromatic pollutants (Wang and Chen 2015).

2D graphene: The main members in this graphene family are GO and rGO, and they possess a similar structure to that of graphene, except that they contain numerous oxygen-derived functional groups on their surface, viz., hydroxyl, carbonyl, and carboxyl groups. These functional groups are introduced via the application of strong oxidants and strong acids. The beneficial aspect of these functional groups in optimising the remediation properties of graphene is governed by adhering additional functional groups or the doping of elements on them, thereby serving as active sites. Besides, GO has a magnificent water dispersion ability, which highly favours the complexation reaction with HM ions.

3d graphene: The major representative groups of 3d graphene are graphene hydrogels and aerogels. As these gels possess large surface area with highly 3D porous framework, it exhibits great adsorption potential and recyclability, thus are of great significance in removal of pollutants from the environment. based upon the data available in the literature, it is estimated that 3D graphene has played pivotal role in removal of cationic dyes such as malachite green (MG), rhodamine B (RhB), methyl blue (MB), fuchsin, methyl violet (MV) due to robust electrostatic interaction owing to their negative charge developed by deprotonation of hydroxyl, carboxyl, and carbonyl functional groups (Xu et al. 2010). Besides, it also exhibits superior adsorption tendency towards HM ions due to electrostatic bonding between HM ions and negatively charged oxygen groups (Cong et al. 2012).

Mesoporous graphene and graphene quantum dots (GQDS): Although these graphene structures do not exhibit high adsorption tendency, their role in environmental remediation is found to be significant. The mesoporous graphene, including Mos_2 and WS_2 with a highly organised porous structure, has gained a lot of attention due to its huge surface area, graphene and mesoporous framework, which confers a congenial environment to the NPs growth and swift charge transfer, simultaneously. These nanocomposites exhibit superior electrochemical generation of hydrogen, excellent electrocatalytic potential, extreme discernible cathodic current, and small overpotential. Quantum dots (QDS) refer 0D nanomaterials, and when graphene sheets with dimensions less than 100 nm are prepared through a chemical tailoring method, the nano-sheets thus obtained display quantum confined electronic attributes and are called GQDS. The GQDS possess an outstanding edge effect, superior biocompatibility, quantum confinement, and can be dispersed in water with great ease. They are of great importance in the photocatalytic degradation of various dyes such as fuchsin (Huang et al. 2018).

2.2.1.1 Adsorption Attributes

Surface area: Graphene is extremely enriched with a large surface area, which reaches up to 2630 m^2/g. This provides the opportunity of having an ample number active sites on the graphene surface, making it a highly suitable material for adsorption of a large number of pollutants (Hsu et al. 2023). The graphene derivatives, viz., GO and rGO, possess better adsorptive removal tendencies as compared to graphene. Graphene oxide (GO) is more hydrophilic, thus its dispersion in the aqueous medium is much higher, which makes it more suitable for the removal of contaminants from the water medium. Moreover, the higher negative charge on the GO surface makes it a much better adsorbent for the remediation of HM ions (Parvathi et al. 2023). Further, GO reduction for preparing rGO leads to the introduction of various defects in its surface, which provides more space for contaminants to get trapped in these voids (Hu et al. 2015). Thus surface area and associated characteristics of graphene and its derivatives play a pivotal role in the adsorption of HM ions.

Surface functional groups: Oxygen (O), nitrogen (N), and sulfur (S)–containing surface functional groups have been recognized as an excellent chelating molecules for robust adsorption of HM ions. It is further observed that graphene-based materials integrated with these atoms-based functional groups exhibit outstanding adsorption performance. Moreover, the O–, N–, and S–based functionalized graphene can be prepared without much hassle (Kong et al. 2021).

(a) *Oxygen (O)-based functionalization:* Graphene structures (GO and rGO) with oxygen-containing functional groups, viz., carboxyl, epoxy, hydroxyl, and carbonyl, are of prime importance in the process of adsorptive removal of HM ions. These functional groups act as the bonding sites for the adsorption of HM ions, evidenced via various functionalization techniques such as XPS (X-ray photoelectron spectroscopy) and FT-IR (Fourier transform infrared spectroscopy) before and after their adsorption (Bian et al. 2015; Pourbeyram 2016). Besides, these functional groups on GO are suitable active sites for developing composites with organic molecules, metals, or various other heteroatoms (Morimoto et al. 2016). With oxygen-based functionalization, GO exhibits much better water absorption capabilities as compared to graphene. The water absorption ability of GO further enhances with increasing mass of GO powder (in grams). This behaviour of GO is attributed to the presence of oxygen-containing functional groups, which act as a site of intercalation which adsorbing the solid particles while allowing the passage of water molecules. Moreover, enhancing the humidity reduced the time to achieve absorption equilibrium, indicating the hydrophilicity of GO (Liu et al. 2017).

(b) *Nitrogen (N)-based functionalization*: The N-based functional groups such as –NH, $-NH_2$, –CONH, –CNH, and –CNR attached to the graphene surface increase its HM ion adsorption capacity. The N atom possesses a lone pair of electrons, which instigates the adsorption of HM ions by serving as an activation site (Ahmad et al. 2020). Zhang and co-workers, via the usage of ethylenediamine, working as a reducing agent and modifier, led to the synthesis of ethylenediamino GO (EDAGO). It exhibited high adsorption ability towards HM ions, including Zn, Ni, Co, and Cu, owing to a large number of hydroxyl and amino groups on their surfaces (Zhang et al. 2018).

(c) *Sulfur (S)-based functionalization*: As the S atom in the functional groups such as –SH, –COSH, –CSSH, L $-CSNH_2$ possesses a lone pair of electrons which provides an excellent site for attracting the HM ions (Ahmad et al. 2020). For instance, via Lewis acid–base reaction, there exists a strong affinity between the

-SH group and numerous HM ions (Pirveysian and Ghiaci 2018). Wei and co-workers fabricated GO functionalized with sulfanilic acid (SGO), which exhibited efficient removal of lead ions (2.00 mmol/g) owing to electrostatic interaction existing between lead ions and S-containing functional groups present on the surface of SGO (Wei et al. 2018).

Electronic properties: The electronic structure of graphene and its derivatives is also a good indicator of HM ion adsorption on their surface. One of the studies reported the interaction and reactivity of GO NPs with neutral and charged lead (Pb) and cadmium (Cd) metal atoms based upon density-functional theory (DFT). It was observed that the charged metal ions were adsorbed more robustly as compared to their neutral counterparts. It was further noticed that the two atoms of the same species (either neutral or charged), when co-adsorbed, exhibited site-specificity of the adsorption energy. Moreover, the binding energy of the singly adsorbed atom was higher than the individual energies of each atom for two co-adsorbed atoms belonging to the same species, indicating a reduction in their adsorption with an increase in their concentration on the adsorption sites of the GO surface. This may be attributed to repulsive behaviour in the case of similarly charged metal ions. Besides, the GO-metal complexes exhibit a little HOMO–LUMO gap (HLG) as compared to GO NPs without metal binding, implying lesser stability of the former compared to the latter. However, all GO-metal complexes possess negative chemical potential, evidencing their stability as well. Thus, deciphering the electronic behaviour of graphene and its derivatives can provide a clear vision to design graphene-based adsorbents for the removal of HM ions (Nianga-Obambi et al. 2024).

2.3 Types of Heavy Metal Adsorption on Graphene Nanomaterials (GNMS)

Graphene and its derivatives, such as GO and rGO, have gained tremendous recognition in adsorptive removal of HM ions, which may be attributed to their unique characteristics, such as wide specific surface area, superior electronic distribution, and innumerable oxygen-containing surface functional groups, as mentioned in the previous section. The interaction of GNMS with HM ions is carried through electrostatic interaction as well as ion exchange. As the lone pair of electrons is owed by the oxygen atom of surface functionalities, which spontaneously form complexes with the HM ions through sharing of electrons, thereby mediating their adsorption on GNMS (Sitko et al. 2021). Graphene NMS exhibit superior adsorption tendency as compared to other conventionally used carbon materials such as activated carbon and carbon nanotubes (CNTS), as reported by various studies. Though the driving force for adsorption mechanism involves electrostatic interaction, complexation, and cation bonding, however, the efficiency of GNMS in adsorption of HM ions depends upon the number and types of oxygen-based surface functionalities as well as the substrate conjugate surface. Electrostatic interaction is the key mechanism for HM ions' adsorption onto the GNMS. As the ph starts to increase gradually, complexes are developed via coordination bond formation between surface hydroxyl groups of GO and HM ions, thereby the key HM ion adsorption mechanism is then taken over by the complex formation and HM ions exchange process (Guo 2023).

Arsenic (As) adsorption: A lot of studies have reported the role of magnetic NPs modified graphene in the removal of inorganic and organic pollutants from the aqueous medium. Water-dispersible magnetite embellished rGO have been observed to play a significant role in the removal of As ions from the polluted water, as reported by Chandra and co-workers (Chandra et al. 2010). The two valence forms of As, viz., As(V) and As(III), respectively, exist in anionic and neutral forms, thus, ph plays a significant role in their adsorption process. It is observed that the adsorbent is capable of adsorbing a large amount of As(V) at low ph, whereas high ph accounts for elevated adsorption of As(III) as the process of complexation occurs immediately after adsorption, inhibiting electrostatic interaction due to increased ph. Further, GO synthesis via the altered Hummer method enhances the surface oxygenated functionalities and enhances the nucleation of the accompanied materials used for functionalization of the graphene. For instance, GO microspheres hybridised with calcium alginate nano-adsorbent are involved in the adsorption of As ions; however, GO is not directly involved in the adsorption process. Nevertheless, it plays a significant role in the fabrication of nucleus-envelope type of adsorbent, which drastically enhances the specific surface area, thereby conferring tremendously high adsorption ability for As(V) and As(III) ions (Basu et al. 2021). Suresh Kumar and co-workers have discovered that single-layer GO and magnetic manganese ferrite nanohybrids (GO-$MnFe_2O4$) exhibit outstanding adsorption characteristics for As and Pb ions (Kumar et al. 2014). The mechanism involved in the adsorption of As ions by GO-$MnFe_2O4$ is dependent upon the response of As and the adsorbent in a varying ph environment. Besides, iron-GO nanohybrids (FeNP@GO) synthesised via the sol–gel method play wonders in the adsorption of As ions with adsorption capacities of 306 mg/g and 431 mg/g, respectively, for As(III) and As(V) (Das et al. 2020). Moreover, the extremely high surface area (252.12 m^2/g) of the obtained nanohybrid materials is an important factor for their elevated As adsorption tendencies. This has particularly reduced the adsorption duration, as it took around only 10 min by FeNP@GO for adsorption of 90% of As. It is thus inferred that oxygen-based surface functionalities in the case of GO have a magnificent impact on the electrostatic and chemical interactions held between As ions and adsorbents. Besides, they mediate the integration of other materials, particularly important for the adsorption process, such as ferric materials, which significantly enhance the adsorption attributes by altering the surface area, porous structure, and surface charge of the adsorbent (Joya-Cárdenas et al. 2022).

The functionalized graphene exhibits extremely high potential for various environmental applications, including adsorption of HM ions (Neri et al. 2015). Engineered or functionalized graphene leads to fine-tuning of the pristine graphene framework and develops different structures (Fig. 2.3a). Both trivalent and pentavalent As ions were subjected to adsorption using GO-iron nanohybrids (GFeN), synthesised via a sol–gel process, without involving the oxidation process of As(III) to As(V) (Das et al.

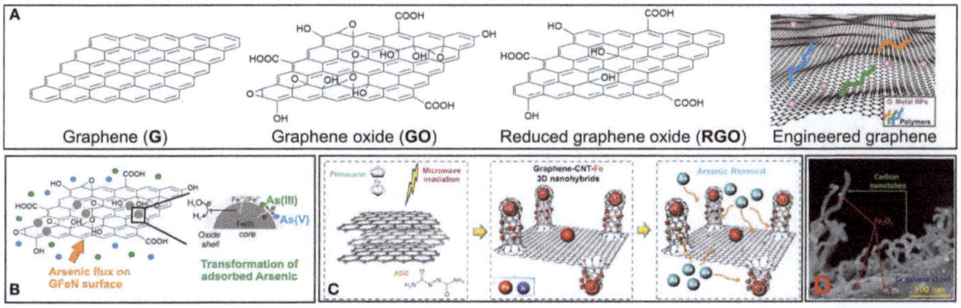

Fig. 2.3 Represents **a** various Graphene-derived materials such as graphene, GO, rGO, and engineered graphene). **b** Illustrative view of GO-iron nanohybrid (GFeN) along with the underlying mechanism for As removal. **c** Illustratio of hierarchically porous 3d assembly (G-CNT-Fe 3d) and the related scanning electron micrographs **d** (Vadahanambi et al. 2013; Foti et al. 2020)

2020). The adsorption of trivalent As ions is mediated through surface complexation. This nanohybrid exhibited great efficiency in the removal of both As ion species, as GO ensures the return of electrons to the Fe NPs by working as an electron reservoir (Fig. 2.3b). Besides, three three-dimensional (3d) graphene structures possess outstanding mechanical stability, thereby being devoid of the aggregation challenge of graphene layers (He et al. 2020). A 3d graphene-based nanohybrid structure, viz., G-CNT-Fe, was prepared, consisting of vertically placed carbon nanotubes (CNT) onto the graphene (G) surface in combination with dispersed iron oxide NPs onto graphene and CNT surface (Vadahanambi et al. 2013) (Fig. 2.3c). This assembly exhibited much higher As(III) species capturing capability as compared to the 2D iron embellished graphene surface. The higher adsorption tendency may be owed to the large surface-to-volume ratio as well as the mesoporous framework, significantly enhancing the molecular diffusion and the reachability of iron oxide particles, acting as As binding sites (Foti et al. 2020).

A novel composite chitosan-GO is an excellent material for adsorption of As(III)/As(V), as evidenced by the elaborated and superior physicochemical attributes (Kumar and Jiang 2016). The ph is an important factor in determining the adsorption efficiency of the adsorbent, as shown in Fig. 2.4. The ph value at which the surface charge density of the adsorbent surface is zero refers to the ph point of zero charges (pHzc). This represents an excellent factor for determining the efficiency of the adsorption mechanism obtained in terms of oppositely charged surfaces. When $pH < pH_{PZC}$, there is a positive charge on the adsorbent surface, which facilitates anion attraction or other positively charged entities. The phpzc is fundamental in directing the process of adsorption of carbon-based NMS by managing the electrostatic attraction between the adsorbate and the adsorbent. The altered electrostatic interaction between As ions and the adsorbent determines the ph influence on the As adsorption mechanism. A plot drawn between the Δph (difference between $pH_{initial}$ and pH_{final}) and $pH_{initial}$ as shown in (Fig. 2.4a) led to the identification of pHzc, and it was in the range of 6.5. Arsenic, both trivalent and pentavalent, species are anionic, thus, positively charged surface enhances electrostatic interaction between them, whereas there is repulsion for negatively charged surfaces (Nata et al. 2011; Xu et al. 2013). It was observed that at low ph, positively charged adsorbents adsorbed As anions. Besides, at this ph, their polar surface functional groups get protonated, and the bulk of them induce a positive charge to the surface, indicating their adsorption tendency is linked with the extent of charge. Hence, it was extracted that adsorbent material with a higher phpc is less efficient for As ions adsorption, whereas with a lower phpc it exhibits enhanced efficiency due to higher positive charge. It is thus obvious that in directing the metal adsorption process, ph is one of the most crucial factors (Fig. 2.4b). Owing to negative charge, ionic species get adsorbed by positively charged adsorbent sites via electrostatic interaction. Alternatively, it could be said that low ph causes adsorbents to work as weak acids and develop positive sites on their surface, facilitating As adsorption through electrostatic attraction (Fig. 2.4c and d). Though electrostatic interaction existing between the adsorbent and the adsorbate for the adsorption process is a dominant force, however, various minor factors such as surface developed metal-owing nodules as well as adsorbate's diffusion into the adsorbents are also credible in this metal uptake mechanism (Fig. 2.4c and d). Figure 2.5 gives the overall view of the adsorption mechanism of As by chitosan-GO adsorbent (Kumar and Jiang 2016).

Chromium (Cr) adsorption: The pristine graphene and its derivatives, such as GO, as adsorbents are of great potential in the removal of chromium metal ions including Cr. Ahmed and co-workers synthesised GO using Hummer's method from graphite powder. The surface characterisation, such as the presence of carboxylic acid

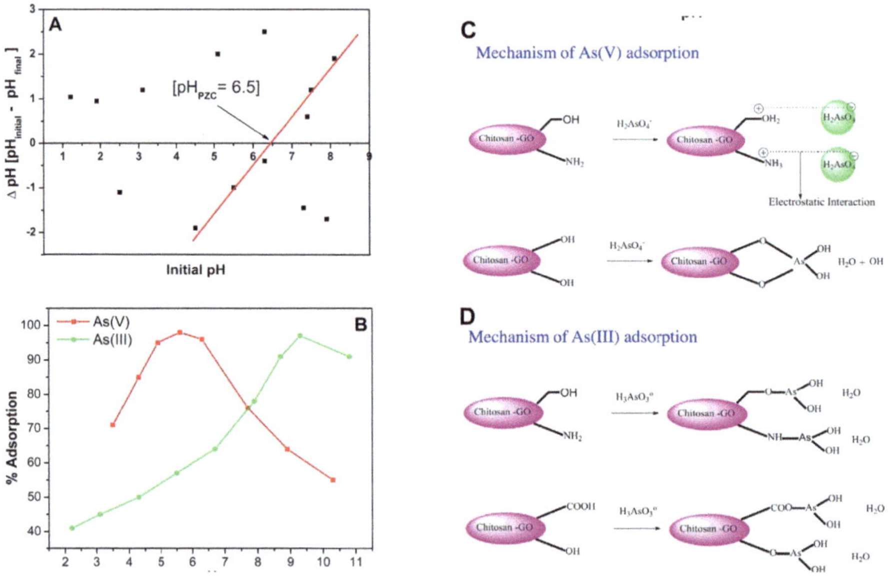

Fig. 2.4 **a** ph point of zero charge (pHzc) **b** Effect of ph **c** mechanism of As(V) adsorption **d** mechanism of As(III) adsorption (Kumar and Jiang 2016)

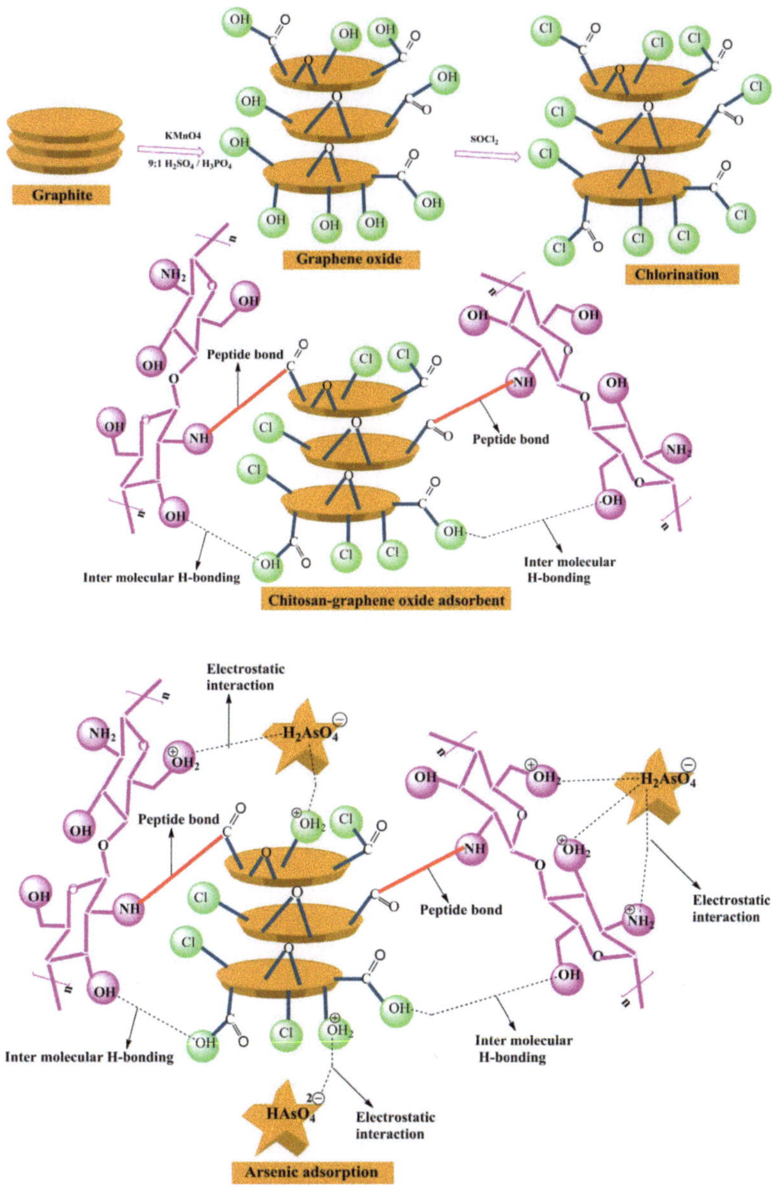

Fig. 2.5 Pictorial representation of interaction mechanism existing between chitosan functionalized-GO observed after adsorption of As(V) (Kumar and Jiang 2016)

groups, was confirmed through characterisation techniques like FT-IR. By measuring the Zeta potential at different pH identified that the pHpcz for the GO surface is close to ph 4.0. Thereafter, the GO was analysed for Cr(III) adsorption mechanism, exhibiting optimal concentration and ph as 0.6 g/L and 4.0, respectively. The process of adsorption occurred quickly and attained the equilibrium within 20 min. The Cr(III) followed the Langmuir and the Freundlich isotherm models and achieved a maximum adsorption of nearly 366 mg/g. GO-mediated removal of tannery effluent was found to be efficient in the case of wastewater with an immense amount of Cr(III) ions and other organic pollutants (Ahmed et al. 2021). Although GO exhibits a great tendency in adsorption of Cr from the medium, the separation of the GO after the adsorption process is difficult and takes a bit longer via the process of centrifugation and filtration. Thereby, it may impose secondary pollution if it is not completely recovered from the environmental medium (Wang et al. 2015). For getting over to this problem, we need to use GO in combination with other NPs such as iron oxide (Fe_3O_4) NPs as it has shown marvellous properties in adsorption of metal ions from aquatic medium owing to their superparamagnetic attributes, lesser Curie temperature, and elevated magnetic coercivity as well as extensive surface area, facile synthesis process, meagre toxicity, and high chemical stability (Bagheri et al. 2017; Rajput et al. 2016). As discussed above in the case of As removal, ph is a fundamental factor for determining the adsorption efficiency of an adsorbent as it impacts numerous events of adsorption, such as the level of ionisation, speciation of the adsorbate, and surface charge of the adsorbent (Mishima et al. 2017). In a study conducted by Jibin et al., it was observed that the maximum ph for efficient adsorption of Cr ions by GO and GO-coated silica NPs (GO/Sio_2) was 3.0 (Jibin et al. 1047). The possible reason is protonation reaction mediated positive charge on GO surface at low ph conditions and existence of Cr(VI) species in the monovalent ($HCrO_4^-$) form converting to divalent species (Cro_4^{2-} and $Cr_2o_7^{2-}$) with increasing ph (Shirkhanloo et al. 2015; Zhang et al. 2015). Thus, positively charged GO and GO/Sio_2 surface attract the negatively charged Cr(VI) species mediated via electrostatic interaction, subsequently enhancing the adsorption process (Matei et al. 2019). In addition, lesser energy is involved in the adsorption of monovalent Cr ions as compared to the divalent form, the monovalent form is more favourably adsorbed. Hence, with increasing ph as the amount of OH^- ions starts dominating, it imparts competition to Cr(VI) ions and decreases in its adsorption rate. The Cr(VI) adsorption rate was found to be nearly 86.155 and 92.8%, respectively, by GO and GO/Sio_2 at 50 ppm concentration. It was further noticed that there was a higher adsorption rate at low metal ion concentration, and a lower adsorption with increased concentration. This is because initially Cr(VI) ion concentration is less compared to the active sites on the adsorbent surface (Samuel et al. 2019). As the concentration enhances, the active available sites get saturated, thus little or no adsorption occurs. As visible from Fig. 2.6, the maximum adsorption rate of Cr(VI) occurred at 50 ppm concentration for both the adsorbents, which drastically declined at 100 and 200 ppm concentration (Jibin et al. 1047).

In another experiment, stable GO was synthesised from graphite via modulated Hummers and Offeman's method (1958) for analysing the process of Cr(VI) adsorption (Mondal and Chakraborty 2020). The most crucial factor in the adsorption

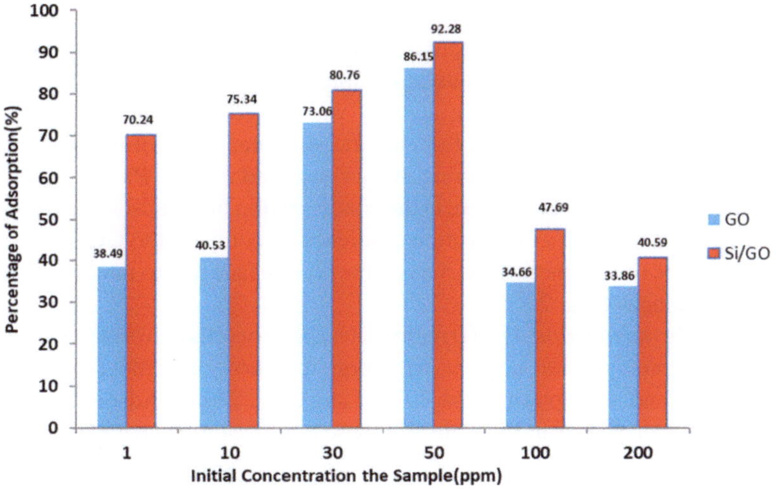

Fig. 2.6 Bar graph representing the adsorption of Cr(VI) using GO and Si/GO at various concentrations (Jibin et al. 1047)

phenomenon is the determination of the phzc of an adsorbent material, and it was found to be 2.134 for GO. As already stated, phpzc refers to the ph at which there is a net negative charge on the adsorbent surface. The ph above this will induce a negative charge on the surface of GO, whereas the value of ph below this will induce a positive charge on its surface (Nayak et al. 2024). The outcome of the study indicated superior potential of GO in capturing the Cr(VI) ions from the aqueous media. At ph 4, the Cr(VI) adsorption on the GO surface is taken over by the monolayer adsorption, exhibiting an adsorption rate of 1.222 mg/g. From analysis of the thermodynamics of the reaction, the adsorption process was found to be endothermic and entropy-driven.

Lead (Pb) adsorption: As graphene is seen to be an effective adsorbent for a variety of organic and inorganic pollutants, it displays magnificent adsorption capabilities for lead as well (Yu et al. 2015). Pristine graphene obtained from graphite is of potential significance in the removal of lead from polluted environments. As graphite is an expensive carbon source, an economic way of synthesising GO, i.e., from waste material, has been devised (Akhavan et al. 2014). Azam and co-workers obtained GO by using graphite rods from disposed dry cell batteries and examined its Pb adsorption efficiency.

The Pb(II) adsorption output of GO, when analysed, was observed to there was nearly 98.87% removal rate of Pb(II) (10 ppm concentration) at ph 4, an adsorbent amount of 0.25 g/L, and a contact time of 20 min. The adsorption maximum was found to be 55.89 mg/g at a temperature of 20°°c. An electrostatic force exists between the cationic HM ions, such as Pb, and the GO surface as opposite charges attract each other (Fig. 2.7a). Moreover, the adsorption mechanism is governed by the FTIR spectra of graphite and GO before and after the binding of lead metal ions (Fig. 2.7b) (Azam et al. 2022). Al-Yaari and Saleh developed an efficient and unconventional adsorbent for the removal of highly obnoxious lead metal ions by grafting GO and polyamide (polyethyleneimine) (PEI) (Al-Yaari and Saleh 2023). The enhanced Pb ions adsorption efficiency was linked with the additional active surface charges of GO as well as the elaborated contact area of the adsorbent on integration with the polyethyleneimine. It was observed that the Pb ion removal rate increased drastically with an increase in the concentration of the adsorbent. Moreover, the GO/PEI exhibited higher adsorption efficiency as compared to GO alone, owing to the appearance of polymeric branches on the GO surface due to integration of PEI, seen in SEM images (Fig. 2.8).

Mercury (Hg) adsorption: Hg(II) removal has been practised in several ways, such as coagulation, precipitation, solvent extraction, membrane filtration, ion-exchange, photocatalysis, and adsorption (Pohl 2020; Ling et al. 2015). Adsorption amongst these is highly appropriate and exhibits immense promise owing to greater efficiency, low operational cost, facile technique, and does not require a secondary treatment process (Liu et al. 2020; Al-Ghouti et al. 2019). For adsorption of Hg ions, various traditional materials such as zeolites, resins, clays, activated carbon, mesoporous carbon and silica have been used, but exhibited poor efficiency and low affinity for Hg(II) ions. Besides, materials with large surface area, viz., metal organic frameworks (MOFS) and functionalized organic polymers, possessed greater effectiveness in adsorption of Hg ions, however, these adsorbents exhibit low stability in aqueous media (Zhao et al. 2021). In this regard, reduced GO (rGO) has been identified as an effective adsorbent for HM ions, including Hg(II) ions, as it possesses high biocompatibility, low cost for preparation, and hydrophilic attributes (30). Tene et al. reported the removal of Hg(II) ions using a green-synthesised rGO, which exhibited nearly 75% removal efficiency for these ions within 20 min, as compared to other conventional adsorbents prepared for Hg(II) ions removal (Tene et al. 2022). Further, rGO functionalization with thiol functional groups facilitates the selectivity and uptake tendency of the adsorbent (Li et al. 2014; Bao et al. 2021). Besides, the challenges with conventional Hg(II) ion adsorbents, such as structural instability in different temperature and ph ranges, can also be addressed using as-prepared rGO.

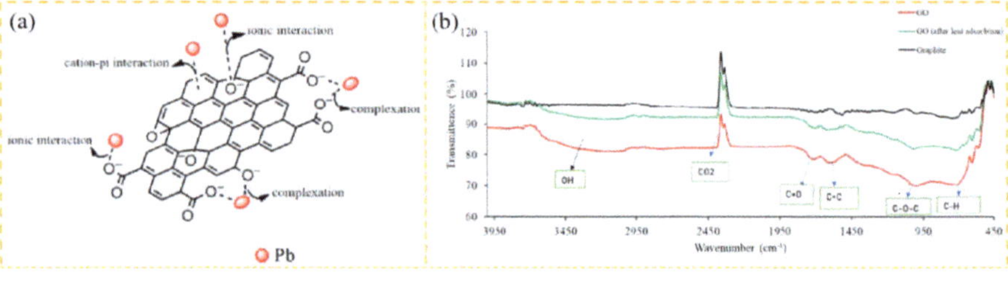

Fig. 2.7 **a** The predicted adsorption process of Pb(II) onto the GO surface in an aqueous media. **b** FTIR spectrum of graphite powder, GO, and lead-adsorbed GO (Azam et al. 2022; Li et al. 2024)

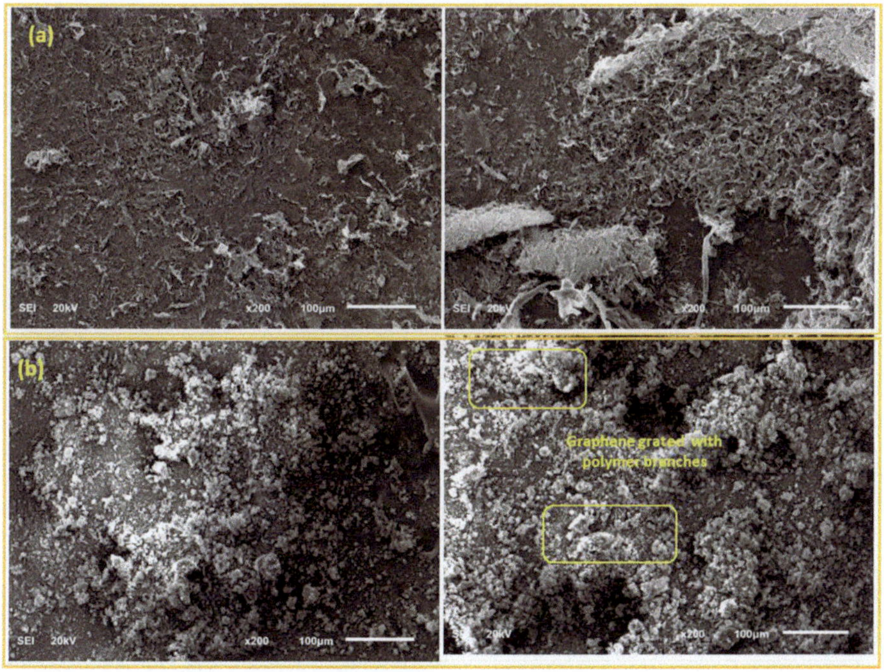

Fig. 2.8 Represents the scanning electron micrographs of **a** graphene oxide and **b** PEI/GO (Al-Yaari and Saleh 2023)

Cadmium (Cd) adsorption: Several studies have been conducted for the removal of another toxic metal ion, namely Cd(II), using GO and rGO. Both these graphene-based adsorbents exhibit high potential in the removal of Cd metal ions, nearly as high as 530 mg/g, when compared to conventional adsorbent materials (Sitko et al. 2013). Bian and co-workers investigated the role of oxygen-based functional groups attached to the GO surface in the adsorption process of HM ions. It was observed that the phenolic hydroxyl and carboxyl groups attached to the GO surface are involved in the adsorption mechanism as depicted by XPS and FTIR spectra (Bian et al. 2015). The optimal ph for efficient adsorption of Cd ions by GO was found to exist between 6–7, mediated by electrostatic force between them. Further, ph beyond this limit enhanced the repulsive force between the Cd ions and negatively charged GO surface functional groups, thereby declining the adsorption process.

Figure 2.9 represents the atomic force microscopy images of GO before and after adsorption of Cd ions. The thickness of GO nanosheets was found to be nearly 1.9 nm indicating bilayer structure as single layer thickness of GO nanosheets is nearly 0.8–1.0 nm (Zhang et al. 2010). However, on adsorption of Cd(II) ions, a lot of projections were seen as shown in the Fig. 2.9c and d) on the GO surface.

Copper (Cu) adsorption: Copper, being an indispensable material for microelectronics, is used extensively, enhancing its accumulation in the environment. Therefore, a lot of work has been conducted on the application of graphene-based materials

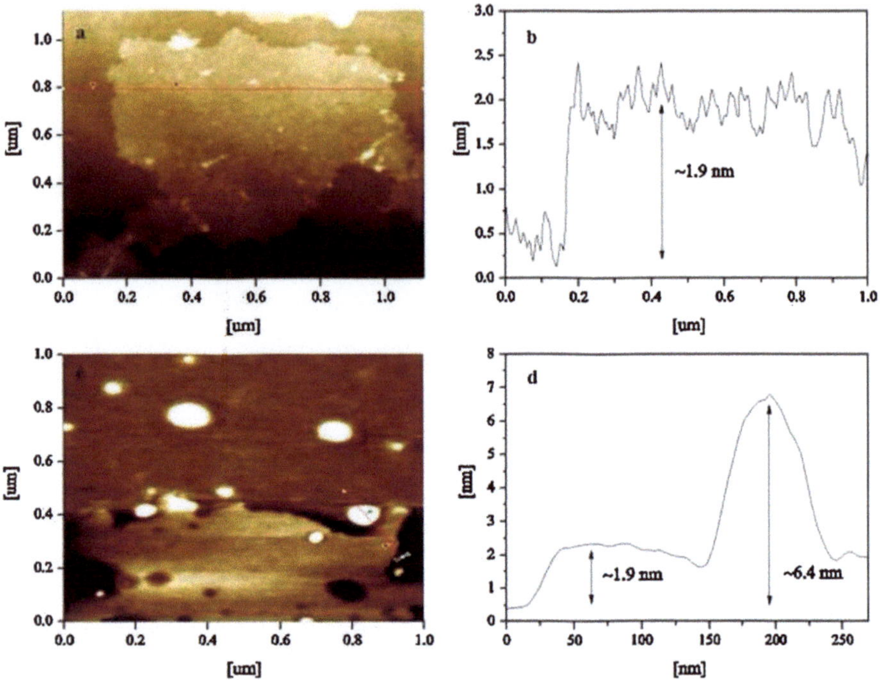

Fig. 2.9 Distinctive AFM images of GO prior to (**a**) and afterwards (**c**) the adsorption process and respective cross-section graphical analysis (**b**) and (**d**) (Bian et al. 2015)

for the removal of copper ions. The most commonly used graphene NMS includes GO membranes/sheets, magnetic GO sheets, and GO functionalized with amino groups for Cu ions adsorption. The amino groups confer additional binding sites on the GO surface for metal ions. The forces involved in the adsorption of Cu ions on the GO surface are electrostatic attraction, ion-exchange, and surface complexation (Krishna et al. 2023).

2.3.1 *Graphene-Based Adsorption Mechanism*

In the process of adsorption, the basic terminologies have been mentioned in Fig. 2.10. The material or any solid substance which provides a surface for binding the HM ions is termed as an adsorbent, and the substance or metal ions which get adsorbed onto its surface are called adsorbate. The alterations in the attributes of the liquid phase, including temperature, ph, and dose, allow the adsorbed species can be released from the surface of the adsorbent, termed as desorption. The adsorbent surface area and porosity play a critical role as the adsorption is a surface phenomenon (Tran et al. 2017). The process of adsorption can be monolayer or multilayer. Monolayer

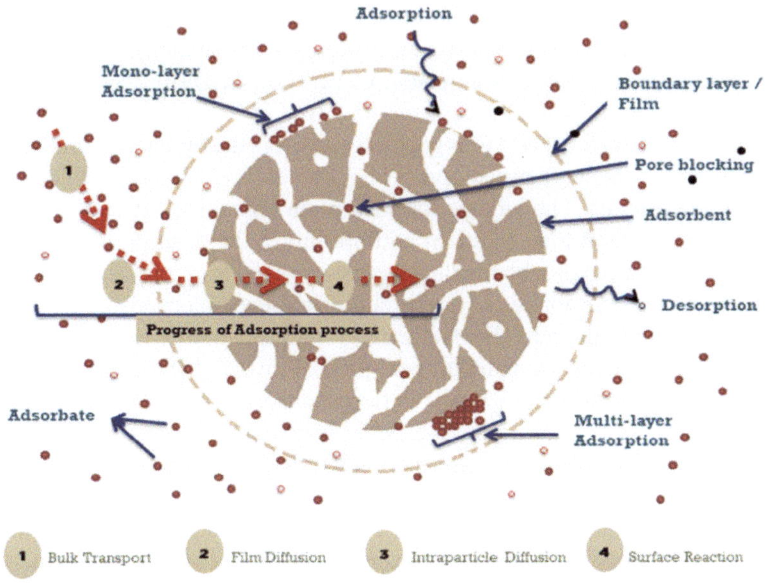

Fig. 2.10 Pictorial view of different terminologies used in the adsorption mechnaim (Krishna et al. 2023)

adsorption is prevalent in the case of chemisorption, and multilayered adsorption is found in the case of physisorption (Ghorbani et al. 2020).

Exploring the mechanism for HM ions adsorption is an important factor for enhancing the adsorption efficiency of an adsorbent. It is obtained from various studies that heteroatom-derived surface functional groups are of immense importance in governing the adsorption process of HM ions. Moreover, the process of adsorption is not a single route mechanism, rather, it is influenced by diverse pathways, either directly or indirectly. For instance, oxygen-derived functional groups in power with sulphur-based functional moieties provide robust propulsion for HM ions adsorption through electrostatic interaction and hydrogen bond development with HM ions, in particular cationic ones. Graphene or its derivatives involve three key mechanisms for adsorption of HM ions, viz., adsorption isotherms, adsorption kinetics, and adsorption thermodynamics, detailed.

Adsorption isotherm: It is a crucial parameter for determining the adsorbent and adsorbate's interaction as well as output efficiency of an adsorbent by applying isotherm models at an equilibrium point and constant temperature (Al-Ghouti and Da'ana 2020). Adsorption isotherms refer to the curves obtained with adsorption functionalities such as adsorbate content accumulated on the adsorbent's surface and the equilibrium dosage of the adsorbate at a fixed time in a particular solution. It can also be obtained by drawing a curve between the same factors but at a constant temperature. Adsorbate's equilibrium concentration is generally represented as $q_{e,}$ and the adsorbate's concentration on the surface of the adsorbent is represented as C_e

(Sahu and Singh 2019). Depending upon the different parameters, various adsorption isotherm models have been proposed, such as Henry's, Freundlich, Langmuir, Hasley, Dubinin-Radushkevich, Redlich-Peterson, Weber-Van VlietFritz Schlunder isotherm models, categorised on several parameters involved (Al-Ghouti and Da'ana, 2020). Graphene-based removal of HM ions generally involves the Freundlich and Langmuir isotherm models, which are based upon two parameters (Zhao et al. 2016).

Adsorption kinetics: The models for adsorption kinetics are meant to determine the adsorption rate and time to attain equilibrium in the adsorption process. This involves the exploration of whether the reaction involved is of pseudo, first, or second order. These models are applied to interpret the time-dependent adsorption tendency of the adsorbate onto the adsorbent surface (Tsibranska and Hristova 2010). These models determine the diverse mass transfer process involved in the adsorption mechanism as well as its rate and efficiency. Different adsorption kinetic models involved in governing the adsorption rate have been developed, viz., pseudo-first, second order, mixed order, n^{th} order models, and mass transfer models based upon intrinsic and extrinsic diffusion (Wang et al. 2021). The different diffusion models linked with adsorption of HM ions or other solid particles include the Weber and Morris as well as the Boyd model, respectively, for intra-particle diffusion in solution systems and intra-particle diffusion of spherical components along with adsorption kinetics determination (Tsibranska and Hristova 2010).

Wang et al. studied adsorptive behaviour of rGO towards Pb(II), Cu(II), Mn(II), and Cd(II) ions and reported their adsorptive kinetics and isotherm, discussed below (Wang et al. 2014).

Adsorptive kinetics: The adsorption of given HM ions onto RGO was enhanced significantly within 30 min of treatment, thereafter it declined and reached equilibrium in an hour. Attaining the adsorptive equilibrium in such a short period indicates high adsorption output of rGO suitable for industrial applications. For analysing the kinetic attributes of the adsorption process, different kinetic models viz., the pseudo-first order and pseudo-second order have been implicated for this experimental work (Ho and McKay 1998). Equations (2.1) and (2.2) respectively are the pseudo-first order and pseudo-second order kinetic models, where Q_e and Q_t represent the amount of HM ions adsorbed (mg/g), correspondingly, at equilibrium and at contact time t (min) and K_1 (1/min) is a rate constant.

$$-\boldsymbol{ln}\left(1-\frac{\boldsymbol{Q}_t}{\boldsymbol{Q}_e}\right)=\boldsymbol{k}_1\boldsymbol{t}+\boldsymbol{C},\boldsymbol{F}={}^{\boldsymbol{Q}_t}\!/_{\boldsymbol{Q}_e} \tag{2.1}$$

$$\frac{\boldsymbol{t}}{\boldsymbol{Q}_t}=\frac{1}{\boldsymbol{K}_2\boldsymbol{Q}_e^2}+\left(\frac{1}{\boldsymbol{Q}_e}\right)\boldsymbol{t} \tag{2.2}$$

According to pseudo-first order kinetic model, the R^2 values for Pb(II), Cu(II), Mn(II), and Cd(II) were found to be 0.8848, 0.8947, 0.9298, and 0.9853, respectively, whereas it was found to be 0.9999, 0.9965, 0.9991, and 0.9999 with pseudo-second order kinetic model. It is thus interpreted that the pseudo-second order kinetic model represented a much better interrelation as compared to the pseudo-first order kinetic model for adsorption of Pb(II), Cu(II), Mn(II), and Cd(II) ions onto rGO. It is then assumed that adsorption of HM ions in this experiment followed pseudo-second order type reaction kinetics. The basis of this model is assuming the rate-determining step involves chemical adsorption, viz., valence forces originating via sharing of electrons existing between adsorbate and adsorbent (Bai et al. 2011; Taty-Costodes et al. 2003).

Adsorption isotherm: with increase in the initial dosage of HM ions ranging from 1 to 6 mmol/L, there was increase in the adsorption ability of rGO for Pb(II), Cu(II), Mn(II), and Cd(II) ions respectively in the range 173.41–396-63, 38.94–54.17, 22.97–38.61, and 41.11–115.25, at 25 °C. Thus, the adsorption tendency of rGO followed the trend Pb(II) > Cd(II) > Cu(II) > Mn(II).

The adsorption data for HM ions using rGO were interpreted via the Langmuir and the Freundlich adsorption isotherms models.

Equation (2.3) and (2.4) represent the Langmuir adsorption equation, wherein Q_m (mg/g) indicates the highest adsorption ability of HM ion by an adsorbent (per unit weight); b is the equilibrium constant (L/mg) of the adsorption process (Langmuir 1918).

$$Q_e = Q_m \frac{bC_e}{1 + bC_e} \tag{2.3}$$

$$\frac{C_e}{Q_e} = \frac{C_e}{Q_m} + \frac{1}{bQ_m} \tag{2.4}$$

Equations (2.5) and (2.6) represent the Freundlich adsorption equation, wherein k and n indicate Freundlich constants. k highlights the adsorption ability (mg/g) and 1/n gives the adsorption intensity.

$$Q_e = kC_e^{1/n} \tag{2.5}$$

$$lnQ_e = lnk + \frac{1}{n} lnC_e \tag{2.6}$$

When data from both the equations were compared, it was seen that R^2 values, respectively for Pb(II), Cu(II), Mn(II), and Cd(II) ions, obtained from the Langmuir adsorption Eq. (0.9976, 0.9994, 0.9964, and 0.9868) fitted well as compared to the Freundlich adsorption Eq. (0.7823, 0.8119, 0.9810, and 0.9810) for experimental equilibrium data. Moreover, as R^2 values were closer to 1, it was inferred that HM ions adsorption via rGO was a peculiar monomolecular layer adsorption process.

Table 2.1 summarises the adsorption isotherm and adsorption kinetics involved in the HM ion adsorption through graphene NMS reported in different studies.

Forces involved in the adsorption of HM ions: All the graphene-based HM ions adsorbents including graphene-derived monolayer adsorbents, graphene-derived multilayer adsorbents, or structurally modified graphene-based adsorbents, and graphene-derived composite structures, mediates the process of adsorption via complexation and reduction existing between HM ions and functional materials/

Table 2.1 Represents the inventory of various metals studies which exhibits the fitted isotherm and kinetic models (Nirmala et al. 2022)

Heavy metal	Graphene-based adsorbent	Interaction mechanism	Adsorption isotherm	Kinetic model	References
Chromium (VI)	Graphene oxide/Chitosan aerosol microspheres	Complexation coordination, electrostatic interaction	Freundlich	Pseudo-second order	Yu et al. (2017)
	Amino-modified Graphene oxide	Electrostatic interaction	Langmuir	Pseudo-second order	Zhao et al. (2016)
Copper (Cu II)	Chitosan/Graphene oxide monolith		Langmuir	Pseudo-second order	Wang et al. (2014)
	Graphene oxide/Chitosan aerosol microspheres	Chelation, electrostatic interaction	Freundlich	Pseudo-first order	Yu et al. (2017)
	Triethylenetetramine-magnetite reduced Graphene oxide		Langmuir	Pseudo-second order	Chen et al. (2015)
Lead (II)	Graphene oxide/Chitosan aerosol microspheres	Chelation, electrostatic interaction	Freundlich (high temp.), Langmuir (low temp.)	Pseudo-second order	Yu et al. (2017)
	Glycol-Graphene oxide-aerogel	Electrostatic interaction	Freundlich	Elovich	Fang et al. (2017)
	Sodium alginate/Graphene oxide aerogel		Langmuir	Pseudo-second order	Jiao et al. (2016)
Mercury (II)	Chitosan-Magnetic Graphene oxide		Langmuir	Pseudo-second order	Zhang et al. (2014)
	Xanthate		Langmuir	Pseudo-second order	Cui et al. (2015)
Arsenic (III)	Activated carbon Graphene oxide	Chemisorption	Langmuir		Chen et al. (2015)
Arsenic (V)	Magnesium Graphene oxide	Chemisorption	Langmuir	Pseudo-second order	Kang et al. (2017)

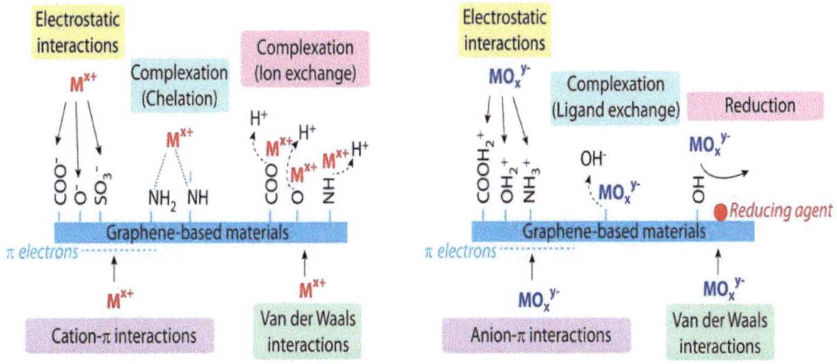

Fig. 2.11 Schematic representation of proposed adsorption mechanisms for removal of HMIs via graphene-derived adsorbents (Wang et al. 2021)

groups, electrostatic attraction, Van der Waals forces, and interaction (Fig. 2.11) (Wang et al. 2021). The interaction generally exists between HM ions and delocalized electrons dissipated on the graphene plane of graphene-derived adsorbents (Wang et al. 2014).

Electrostatic interaction: The oxygen-based surface functional groups impart a negative charge to GO-based adsorbents, as these are present in abundance. These negatively charged adsorbents exhibit great affinity for positively charged HM ions. Moreover, under basic ph conditions, the deprotonation of carboxyl groups attached to the GO surface is enhanced, thereby further elevating their output in adsorption of oppositely charged HM ions (Xu et al. 2013). However, these interactions are not favourable for negatively charged HM ions; rather, they show a repulsive force. However, to enhance the adsorption capacity of graphene-based adsorbents, it must be reduced or altered by integrating it with amino groups. The electrostatic interactions are strong, thereby hampering the process of desorption, which further impacts the process of adsorption. Thus, fine-tuning is needed in the structure of adsorbents, which may confer weak electrostatic interaction without impacting the adsorption ability (Xu et al. 2013).

π–π stacking: The πinteractions are found between the graphene adsorbents and the aromatic rings (Yang et al. 2008). Different computational and experimental studies reported the existence of these interactions between the sp^2 carbon atoms of the adsorbent and aromatic pollutants. Therefore, πstacking is of prime importance for adsorptive removal of organic pollutants, and they are less involved in the adsorption of inorganic pollutants, including HM ions (Yang et al. 2008).

Hydrophobic interaction: These interaction also contributes significantly to the adsorptive removal of pollutants using graphene-based adsorbents. The major force involved in driving hydrophobic interactions is an entropic phenomenon generated via eliminating the ordered water molecules from the nonpolar interface. The hydrophobic pollutants are adsorbed and removed through hydrophobic interaction existing between them and graphene structure. Besides, hydrophilic pollutants containing some hydrophobic chain or part are also eliminated efficiently via these interactions (Xu et al. 2013).

Coordination: Though coordination has less involvement in the process of adsorption, HM ions adsorption using GO and its derivatives involves coordination as it develops between oxygen atoms and HM ions. A study compared the folding of GO sheets with copper ions and sodium (Na) ions (exhibiting the same ionic strength) (Yang et al. 2010). It was observed that Cu ions mediated strong folding of GO sheets as compared to Na ions. This coordination-driven spectrum shift of Cu ions was also evidenced through UV–VIS spectra. Hence, in the remediation process of HM ions via adsorption, coordination must be taken into account and even designed. In this regard, Madadrang and co-workers fabricated GO in linkage with EDTA via the silylation technique. This EDTA coordinated with several HM ions, including Pb, Cu, Cd, Hg, and Ni and was chelated through EDTA groups (Madadrang et al. 2012).

2.4 Conclusions

Heavy metal remediation through adsorption is a fundamental process mediated by precipitation, ion-exchange, electrostatic interaction, surface complexation, hydrophobic interaction, and stacking. Graphene and its derivatives, owing to their superior physicochemical properties such as large surface-to-volume ratio, robust mechanical attributes, unique electronic attributes, and diverse surface functionalities, are gaining immense importance as an adsorbent for the removal of HM ions from the environment. The current chapter describes the role, mechanism and advantages of graphene and its derivatives for the process of adsorptive removal of HM ions such as Pb, Cr, As, Hg, Cu, Cd, etc.

References

Ahmad SZN, Wan Salleh WN, Ismail AF, Yusof N, Mohd Yusop MZ, Aziz F (2020) Adsorptive removal of heavy metal ions using graphene-based nanomaterials: Toxicity, roles of functional groups and mechanisms. Chemosphere 248:126008. https://doi.org/10.1016/j.chemosphere.2020.126008

Ahmed S, Fatema-Tuj-Zohra, Mahdi MM, Mahmudunnabi DM, Choudhury TR, Alam MZ, Nurnabi M (2021) Synthesis and characterization of graphene oxide for removal of Cr(III) from tannery effluent. Desalination Water Treat 244:201–211. https://doi.org/10.5004/dwt.2021.27895

Akhavan O, Bijanzad K, Mirsepah A (2014) Synthesis of graphene from natural and industrial carbonaceous wastes. RSC Adv 4(39):20441. https://doi.org/10.1039/c4ra01550a

Al-Ghouti MA, Da'ana DA (2020) Guidelines for the use and interpretation of adsorption isotherm models: A review. J Hazard Mater, 393:122383. https://doi.org/10.1016/j.jhazmat.2020.122383

Al-Ghouti MA, Da'ana D, Abu-Dieyeh M, Khraisheh M (2019) Adsorptive removal of mercury from water by adsorbents derived from date pits. Sci Rep 9(1):15327. https://doi.org/10.1038/s41598-019-51594-y

Al-Yaari M, Saleh TA (2023) Removal of lead from wastewater using synthesized polyethyleneimine-grafted graphene oxide. Nanomaterials 13(6):1078. https://doi.org/10.3390/nano13061078

Azam MG, Kabir MH, Shaikh MAA, Ahmed S, Mahmud M, Yasmin S (2022) A rapid and efficient adsorptive removal of lead from water using graphene oxide prepared from waste dry cell battery. J Water Process Eng 46:102597. https://doi.org/10.1016/j.jwpe.2022.102597

Bagheri AR, Ghaedi M, Asfaram A, Bazrafshan AA, Jannesar R (2017) Comparative study on ultrasonic assisted adsorption of dyes from single system onto Fe_3O_4 magnetite nanoparticles loaded on activated carbon: Experimental design methodology. Ultrason Sonochem 34:294–304. https://doi.org/10.1016/j.ultsonch.2016.05.047

Bai L, Hu H, Fu W, Wan J, Cheng X, Zhuge L, Xiong L, Chen Q (2011) Synthesis of a novel silica-supported dithiocarbamate adsorbent and its properties for the removal of heavy metal ions. J Hazard Mater 195:261–275. https://doi.org/10.1016/j.jhazmat.2011.08.038

Bao S, Wang Y, Yu Y, Yang W, Sun Y (2021) Cross-linked sulfydryl-functionalized graphene oxide as ultra-high capacity adsorbent for high selectivity and ppb level removal of mercury from water under wide pH range. Environ Pollut 271:116378. https://doi.org/10.1016/j.envpol.2020.116378

Basu H, Singh S, Venkatesh M, Pimple MV, Singhal RK (2021) Graphene oxide-MnO_2-goethite microsphere impregnated alginate: A novel hybrid nanosorbent for As (III) and As (V) removal from groundwater. J Water Process Eng 42:102129. https://doi.org/10.1016/j.jwpe.2021.102129

Bian Y, Bian Z-Y, Zhang J-X, Ding A-Z, Liu S-L, Wang H (2015) Effect of the oxygen-containing functional group of graphene oxide on the aqueous cadmium ions removal. Appl Surf Sci 329:269–275. https://doi.org/10.1016/j.apsusc.2014.12.090

Chandra V, Park J, Chun Y, Lee JW, Hwang I-C, Kim KS (2010) Water-dispersible magnetite-reduced graphene oxide composites for arsenic removal. ACS Nano 4(7):3979–3986. https://doi.org/10.1021/nn1008897

Chen JH, Xing HT, Sun X, Su ZB, Huang YH, Weng W, Hu SR, Guo HX, Wu WB, He YS (2015) Highly effective removal of Cu(II) by triethylenetetramine-magnetic reduced graphene oxide composite. Appl Surf Sci 356:355–363. https://doi.org/10.1016/j.apsusc.2015.08.076

Chen M-L, Sun Y, Huo C-B, Liu C, Wang J-H (2015) Akaganeite decorated graphene oxide composite for arsenic adsorption/removal and its proconcentration at ultra-trace level. Chemosphere 130:52–58. https://doi.org/10.1016/j.chemosphere.2015.02.046

Cong H-P, Ren X-C, Wang P, Yu S-H (2012) Macroscopic multifunctional graphene-based hydrogels and aerogels by a metal ion induced self-assembly process. ACS Nano 6(3):2693–2703. https://doi.org/10.1021/nn300082k

Cui L, Guo X, Wei Q, Wang Y, Gao L, Yan L, Yan T, Du B (2015) Removal of mercury and methylene blue from aqueous solution by xanthate functionalized magnetic graphene oxide: Sorption kinetic and uptake mechanism. J Colloid Interface Sci 439:112–120. https://doi.org/10.1016/j.jcis.2014.10.019

Cui L, Wang Y, Gao L, Hu L, Yan L, Wei Q, Du B (2015) EDTA functionalized magnetic graphene oxide for removal of Pb(II), Hg(II) and Cu(II) in water treatment: Adsorption mechanism and separation property. Chem Eng J 281:1–10. https://doi.org/10.1016/j.cej.2015.06.043

Das TK, Sakthivel TS, Jeyaranjan A, Seal S, Bezbaruah AN (2020) Ultra-high arsenic adsorption by graphene oxide iron nanohybrid: Removal mechanisms and potential applications. Chemosphere 253:126702. https://doi.org/10.1016/j.chemosphere.2020.126702

Dong L, Hou L, Wang Z, Gu P, Chen G, Jiang R (2018) A new function of spent activated carbon in BAC process: Removing heavy metals by ion exchange mechanism. J Hazard Mater 359:76–84. https://doi.org/10.1016/j.jhazmat.2018.07.030

Fang Q, Zhou X, Deng W, Liu Z (2017) Hydroxyl-containing organic molecule induced self-assembly of porous graphene monoliths with high structural stability and recycle performance for heavy metal removal. Chem Eng J 308:1001–1009. https://doi.org/10.1016/j.cej.2016.09.139

Fomkin AA, Men'shchikov IE, Pribylov AA, Gur'yanov VV, Shkolin AV, Zaitsev DS, Tvardovskii AV (2017) Methane adsorption on microporous carbon adsorbent with wide pore size distribution. Colloid Journal 79(1):144–151. https://doi.org/10.1134/S1061933X16060053

Foti C, Mineo PG, Nicosia A, Scala A, Neri G, Piperno A (2020) Recent advances of graphene-based strategies for arsenic remediation. Front Chem 8. https://doi.org/10.3389/fchem.2020.608236

Fu R, Liu Y, Lou Z, Wang Z, Baig SA, Xu X (2016) Adsorptive removal of Pb(II) by magnetic activated carbon incorporated with amino groups from aqueous solutions. J Taiwan Inst Chem Eng 62:247–258. https://doi.org/10.1016/j.jtice.2016.02.012

Ghorbani M, Seyedin O, Aghamohammadhassan M (2020) Adsorptive removal of lead (II) ion from water and wastewater media using carbon-based nanomaterials as unique sorbents: A review. J Environ Manage 254:109814. https://doi.org/10.1016/j.jenvman.2019.109814

Guo T (2023) Application of graphene technology in the removal of heavy metal ions in wastewater. Highlights Sci, Eng Technol 73:464–470. https://doi.org/10.54097/hset.v73i.14053

He J, Ni F, Cui A, Chen X, Deng S, Shen F, Huang C, Yang G, Song C, Zhang J, Tian D, Long L, Zhu Y, Luo L (2020) New insight into adsorption and co-adsorption of arsenic and tetracycline using a Y-immobilized graphene oxide-alginate hydrogel: Adsorption behaviours and mechanisms. Sci Total Environ 701:134363. https://doi.org/10.1016/j.scitotenv.2019.134363

Ho YS, McKay G (1998) Sorption of dye from aqueous solution by peat. Chem Eng J 70(2):115–124. https://doi.org/10.1016/S0923-0467(98)00076-1

Hsu C-Y, Rheima AM, Mohammed MS, Kadhim MM, Mohammed SH, Abbas FH, Abed ZT, Mahdi ZM, Abbas ZS, Hachim SK, Ali FK, Mahmoud ZH, Kianfar E (2023) Application of carbon nanotubes and graphene-based nanoadsorbents in water treatment. Bionanoscience 13(4):1418–1436. https://doi.org/10.1007/s12668-023-01175-1

Hu Y, Song S, Lopez-Valdivieso A (2015) Effects of oxidation on the defect of reduced graphene oxides in graphene preparation. J Colloid Interface Sci 450:68–73. https://doi.org/10.1016/j.jcis.2015.02.059

Huang B, He J, Bian S, Zhou C, Li Z, Xi F, Liu J, Dong X (2018) S-doped graphene quantum dots as nanophotocatalyst for visible light degradation. Chin Chem Lett 29(11):1698–1701. https://doi.org/10.1016/j.cclet.2018.01.004

Inyang M, Gao B, Ding W, Pullammanappallil P, Zimmerman AR, Cao X (2011) Enhanced lead sorption by biochar derived from anaerobically digested sugarcane bagasse. Sep Sci Technol 46(12):1950–1956. https://doi.org/10.1080/01496395.2011.584604

Inyang M, Gao B, Yao Y, Xue Y, Zimmerman AR, Pullammanappallil P, Cao X (2012) Removal of heavy metals from aqueous solution by biochars derived from anaerobically digested biomass. Bioresour Technol 110:50–56. https://doi.org/10.1016/j.biortech.2012.01.072

Inyang MI, Gao B, Yao Y, Xue Y, Zimmerman A, Mosa A, Pullammanappallil P, Ok YS, Cao X (2016) A review of biochar as a low-cost adsorbent for aqueous heavy metal removal. Crit Rev Environ Sci Technol 46(4):406–433. https://doi.org/10.1080/10643389.2015.1096880

Jiao C, Xiong J, Tao J, Xu S, Zhang D, Lin H, Chen Y (2016) Sodium alginate/graphene oxide aerogel with enhanced strength–toughness and its heavy metal adsorption study. Int J Biol Macromol 83:133–141. https://doi.org/10.1016/j.ijbiomac.2015.11.061

Jibin K, Augustine S, Velayudhan P, George J, Krishnageham Sidharthan S, Paulose S, Thomas S (2023) Unleashing the power of graphene-based nanomaterials for chromium(VI) ion elimination from water. Crystals (Basel) 13(7):1047. https://doi.org/10.3390/cryst13071047

Joya-Cárdenas DR, Rodríguez-Caicedo JP, Gallegos-Muñoz A, Zanor GA, Caycedo-García MS, Damian-Ascencio CE, Saldaña-Robles A (2022) Graphene-based adsorbents for arsenic, fluoride, and chromium adsorption: Synthesis methods review. Nanomaterials 12(22):3942. https://doi.org/10.3390/nano12223942

Kang BK, Lim BS, Yoon Y, Kwag SH, Park WK, Song YH, Yang WS, Ahn Y-T, Kang J-W, Yoon DH (2017) Efficient removal of arsenic by strategically designed and layer-by-layer assembled PS@+rGO@GO@Fe_3O_4 composites. J Environ Manage 201:286–293. https://doi.org/10.1016/j.jenvman.2017.05.066

Kong Q, Shi X, Ma W, Zhang F, Yu T, Zhao F, Zhao D, Wei C (2021) Strategies to improve the adsorption properties of graphene-based adsorbent towards heavy metal ions and their compound pollutants: A review. J Hazard Mater 415:125690. https://doi.org/10.1016/j.jhazmat.2021.125690

Krishna RH, Chandraprabha MN, Samrat K, Krishna Murthy TP, Manjunatha C, Kumar SG (2023) Carbon nanotubes and graphene-based materials for adsorptive removal of metal ions—A review on surface functionalization and related adsorption mechanism. Appl Surf Sci Adv 16:100431. https://doi.org/10.1016/j.apsadv.2023.100431

Kumar ASK, Jiang S-J (2016) Chitosan-functionalized graphene oxide: A novel adsorbent an efficient adsorption of arsenic from aqueous solution. J Environ Chem Eng 4(2):1698–1713. https://doi.org/10.1016/j.jece.2016.02.035

Kumar S, Nair RR, Pillai PB, Gupta SN, Iyengar MAR, Sood AK (2014) Graphene oxide–$MnFe_2O_4$ magnetic nanohybrids for efficient removal of lead and arsenic from water. ACS Appl Mater Interfaces 6(20):17426–17436. https://doi.org/10.1021/am504826q

Langmuir I (1918) The adsorption of gases on plane surfaces of glass, mica and platinum. J Am Chem Soc 40(9):1361–1403. https://doi.org/10.1021/ja02242a004

Li B, Zhang Y, Ma D, Shi Z, Ma S (2014) Mercury nano-trap for effective and efficient removal of mercury(II) from aqueous solution. Nat Commun 5(1):5537. https://doi.org/10.1038/ncomms6537

Li G, Du R, Cao Z, Li C, Xue J, Ma X, Wang S (2024) Research progress in graphene-based adsorbents for wastewater treatment: Preparation, adsorption properties and mechanisms for inorganic and organic pollutants. C (Basel) 10(3):78. https://doi.org/10.3390/c10030078

Liang J, Li X, Yu Z, Zeng G, Luo Y, Jiang L, Yang Z, Qian Y, Wu H (2017) Amorphous MnO_2 modified biochar derived from aerobically composted swine manure for adsorption of Pb(II) and Cd(II). ACS Sustain Chem Eng 5(6):5049–5058. https://doi.org/10.1021/acssuschemeng.7b00434

Ling L, Fan M, Wang B, Zhang R (2015) Application of computational chemistry in understanding the mechanisms of mercury removal technologies: A review. Energy Environ Sci 8(11):3109–3133. https://doi.org/10.1039/C5EE02255J

Liu T, Han X, Wang Y, Yan L, Du B, Wei Q, Wei D (2017) Magnetic chitosan/anaerobic granular sludge composite: Synthesis, characterization and application in heavy metal ions removal. J Colloid Interface Sci 508:405–414. https://doi.org/10.1016/j.jcis.2017.08.067

Liu R, Gong T, Zhang K, Lee C (2017) Graphene oxide papers with high water adsorption capacity for air dehumidification. Sci Rep 7(1):9761. https://doi.org/10.1038/s41598-017-09777-y

Liu Z, Liu D, Zhao B, Feng L, Ni M, Jin J (2020) Mercury removal based on adsorption and oxidation by fly ash: A review. Energy Fuels 34(10):11840–11866. https://doi.org/10.1021/acs.energyfuels.0c02209

Lv D, Liu Y, Zhou J, Yang K, Lou Z, Baig SA, Xu X (2018) Application of EDTA-functionalized bamboo activated carbon (BAC) for Pb(II) and Cu(II) removal from aqueous solutions. Appl Surf Sci 428:648–658. https://doi.org/10.1016/j.apsusc.2017.09.151

El kady M, Shokry H, Hamad H (2016) Effect of superparamagnetic nanoparticles on the physicochemical properties of nano hydroxyapatite for groundwater treatment: adsorption mechanism of Fe(<scp>ii</scp>) and Mn(<scp>ii</scp>). RSC Adv 6(85):82244–82259. https://doi.org/10.1039/C6RA14497G

Madadrang CJ, Kim HY, Gao G, Wang N, Zhu J, Feng H, Gorring M, Kasner ML, Hou S (2012) Adsorption behavior of EDTA-graphene oxide for Pb (II) removal. ACS Appl Mater Interfaces 4(3):1186–1193. https://doi.org/10.1021/am201645g

Matei E, Predescu AM, Râpă M, Tarcea C, Pantilimon CM, Favier L, Berbecaru AC, Sohaciu M, Predescu C (2019) Removal of chromium(VI) from aqueous solution using a novel green magnetic nanoparticle—chitosan adsorbent. Anal Lett 52(15):2416–2438. https://doi.org/10.1080/00032719.2019.1601734

Mishima K, Du X, Sekiguchi S, Kano N (2017) Experimental and theoretical studies on the adsorption and desorption mechanisms of chromate ions on cross-linked chitosan. J Funct Biomater 8(4):51. https://doi.org/10.3390/jfb8040051

Mohan D, Pittman CU, Bricka M, Smith F, Yancey B, Mohammad J, Steele PH, Alexandre-Franco MF, Gómez-Serrano V, Gong H (2007) Sorption of arsenic, cadmium, and lead by chars produced from fast pyrolysis of wood and bark during bio-oil production. J Colloid Interface Sci 310(1):57–73. https://doi.org/10.1016/j.jcis.2007.01.020

Mondal NK, Chakraborty S (2020) Adsorption of Cr(VI) from aqueous solution on graphene oxide (GO) prepared from graphite: equilibrium, kinetic and thermodynamic studies. Appl Water Sci 10(2):61. https://doi.org/10.1007/s13201-020-1142-2

Morimoto N, Kubo T, Nishina Y (2016) Tailoring the oxygen content of graphite and reduced graphene oxide for specific applications. Sci Rep 6(1):21715. https://doi.org/10.1038/srep21715

Nata IF, Sureshkumar M, Lee C-K (2011) One-pot preparation of amine-rich magnetite/bacterial cellulose nanocomposite and its application for arsenate removal. RSC Adv 1(4):625. https://doi.org/10.1039/c1ra00153a

Nayak A, Bhushan B, Gupta V, Sharma P (2017) Chemically activated carbon from lignocellulosic wastes for heavy metal wastewater remediation: Effect of activation conditions. J Colloid Interface Sci 493:228–240. https://doi.org/10.1016/j.jcis.2017.01.031

Nayak A, Bhushan B, Kotnala S (2024) Fundamentals and mechanism of adsorption. In: Sustainable technologies for remediation of emerging pollutants from aqueous environment, Elsevier, pp 29–62. https://doi.org/10.1016/B978-0-443-18618-9.00002-4

Neri G, Scala A, Barreca F, Fazio E, Mineo PG, Mazzaglia A, Grassi G, Piperno A (2015) Engineering of carbon based nanomaterials by ring-opening reactions of a reactive azlactone graphene platform. Chem Commun 51(23):4846–4849. https://doi.org/10.1039/C5CC00518C

Nianga-Obambi PS, Douma DH, Etindele AJ, Raji AT, Rodrigue Malonda-Boungou B, M'PASSI-MABIALA B, Kenmoe S (2024) Reactivity and heavy metal adsorption behaviour of graphene oxide nanoparticle from first principle investigations: Case of Pb and CD. https://doi.org/10.20944/preprints202405.0576.v1

Nirmala N, Shriniti V, Aasresha K, Arun J, Gopinath KP, Dawn SS, Sheeladevi A, Priyadharsini P, Birindhadevi K, Chi NTL, Pugazhendhi A (2022) Removal of toxic metals from wastewater environment by graphene-based composites: A review on isotherm and kinetic models, recent trends, challenges and future directions. Sci Total Environ 840:156564. https://doi.org/10.1016/j.scitotenv.2022.156564

Novoselov KS, Fal′ko VI, Colombo L, Gellert PR, Schwab MG, Kim K (2012) A roadmap for graphene. Nature 490(7419):192–200. https://doi.org/10.1038/nature11458

Novoselov KS, Geim AK, Morozov SV, Jiang D, Zhang Y, Dubonos SV, Grigorieva IV, Firsov AA (2004) Electric field effect in atomically thin carbon films. Science (1979), 306(5696):666–669. https://doi.org/10.1126/science.1102896

Pan X, Ji J, Zhang N, Xing M (2020) Research progress of graphene-based nanomaterials for the environmental remediation. Chin Chem Lett 31(6):1462–1473. https://doi.org/10.1016/j.cclet.2019.10.002

Parvathi E, Dilraj N, Akshaya CV, Deepak NK (2023) A review on graphene-based adsorbents for the remediation of toxic heavy metals from aqueous sources. Int J Environ Sci Technol 20(10):11645–11672. https://doi.org/10.1007/s13762-022-04587-w

Peng W, Li H, Liu Y, Song S (2017) A review on heavy metal ions adsorption from water by graphene oxide and its composites. J Mol Liq 230:496–504. https://doi.org/10.1016/j.molliq.2017.01.064

Pirveysian M, Ghiaci M (2018) Synthesis and characterization of sulfur functionalized graphene oxide nanosheets as efficient sorbent for removal of Pb2+, Cd2+, Ni2+ and Zn2+ ions from aqueous solution: A combined thermodynamic and kinetic studies. Appl Surf Sci 428:98–109. https://doi.org/10.1016/j.apsusc.2017.09.105

Pohl A (2020) Removal of heavy metal ions from water and wastewaters by sulfur-containing precipitation agents. Water Air Soil Pollut 231(10):503. https://doi.org/10.1007/s11270-020-04863-w

Pourbeyram S (2016) Effective removal of heavy metals from aqueous solutions by Graphene Oxide-Zirconium Phosphate (GO–Zr-P) nanocomposite. Ind Eng Chem Res 55(19):5608–5617. https://doi.org/10.1021/acs.iecr.6b00728

Qasem NAA, Mohammed RH, Lawal DU (2021) Removal of heavy metal ions from wastewater: A comprehensive and critical review. NPJ Clean Water 4(1):36. https://doi.org/10.1038/s41545-021-00127-0

Rajendran S, Priya AK, Senthil Kumar P, Hoang TKA, Sekar K, Chong KY, Khoo KS, Ng HS, Show PL (2022) A critical and recent developments on adsorption technique for removal of heavy metals from wastewater-A review. Chemosphere 303:135146. https://doi.org/10.1016/j.chemosphere.2022.135146

Raji Z, Karim A, Karam A, Khalloufi S (2023) Adsorption of heavy metals: Mechanisms, kinetics, and applications of various adsorbents in wastewater remediation—a review. Waste 1(3):775–805. https://doi.org/10.3390/waste1030046

Rajput S, Pittman CU, Mohan D (2016) Magnetic magnetite (Fe_3O_4) nanoparticle synthesis and applications for lead (Pb2+) and chromium (Cr6+) removal from water. J Colloid Interface Sci 468:334–346. https://doi.org/10.1016/j.jcis.2015.12.008

Sahu O, Singh N (2019) Significance of bioadsorption process on textile industry wastewater. In: The impact and prospects of green chemistry for textile technology, Elsevier, pp 367–416. https://doi.org/10.1016/B978-0-08-102491-1.00013-7

Samuel MS, Bhattacharya J, Raj S, Santhanam N, Singh H, Pradeep Singh ND (2019) Efficient removal of Chromium(VI) from aqueous solution using chitosan grafted graphene oxide (CS-GO) nanocomposite. Int J Biol Macromol 121:285–292. https://doi.org/10.1016/j.ijbiomac.2018.09.170

Shirkhanloo H, Khaligh A, Mousavi HZ, Rashidi A (2015) Graphene oxide-packed micro-column solid-phase extraction combined with flame atomic absorption spectrometry for determination of lead (II) and nickel (II) in water samples. Int J Environ Anal Chem 95(1):16–32. https://doi.org/10.1080/03067319.2014.983437

Silva Buarque NM, de Brito Buarque HL, Capelo-Neto J (2015) Adsorption kinetics and diffusion of Saxitoxins on granular-activated carbon: Influence of pore size distribution. J Water Supply: Res Technol-Aqua 64(3):344–353. https://doi.org/10.2166/aqua.2015.140

Sitko R, Turek E, Zawisza B, Malicka E, Talik E, Heimann J, Gagor A, Feist B, Wrzalik R (2013) Adsorption of divalent metal ions from aqueous solutions using graphene oxide. Dalton Trans 42(16):5682. https://doi.org/10.1039/c3dt33097d

Sitko R, Musielak M, Serda M, Talik E, Gagor A, Zawisza B, Malecka M (2021) Graphene oxide decorated with fullerenol nanoparticles for highly efficient removal of Pb(II) ions and ultrasensitive detection by total-reflection X-ray fluorescence spectrometry. Sep Purif Technol 277:119450. https://doi.org/10.1016/j.seppur.2021.119450

Taty-Costodes VC, Fauduet H, Porte C, Delacroix A (2003) Removal of Cd(II) and Pb(II) ions, from aqueous solutions, by adsorption onto sawdust of Pinus sylvestris. J Hazard Mater 105(1–3):121–142. https://doi.org/10.1016/j.jhazmat.2003.07.009

Tene T, Arias Arias F, Guevara M, Nuñez A, Villamagua L, Tapia C, Pisarra M, Torres FJ, Caputi LS, Vacacela Gomez C (2022) Removal of mercury(II) from aqueous solution by partially reduced graphene oxide. Sci Rep 12(1):6326. https://doi.org/10.1038/s41598-022-10259-z

Tran HN, You S-J, Hosseini-Bandegharaei A, Chao H-P (2017) Mistakes and inconsistencies regarding adsorption of contaminants from aqueous solutions: A critical review. Water Res 120:88–116. https://doi.org/10.1016/j.watres.2017.04.014

Tsibranska I, Hristova E (2010) Modelling of heavy metal adsorption into activated carbon from apricot stones in fluidized bed. Chem Eng Process 49(10):1122–1127. https://doi.org/10.1016/j.cep.2010.08.003

Vadahanambi S, Lee S-H, Kim W-J, Oh I-K (2013) Arsenic removal from contaminated water using three-dimensional graphene-carbon nanotube-iron oxide nanostructures. Environ Sci Technol:130904083814004. https://doi.org/10.1021/es401389g

Wang J, Chen B (2015) Adsorption and coadsorption of organic pollutants and a heavy metal by graphene oxide and reduced graphene materials. Chem Eng J 281:379–388. https://doi.org/10.1016/j.cej.2015.06.102

Wang B, Zhang F, He S, Huang F, Peng Z (2014) Adsorption behaviour of reduced graphene oxide for removal of heavy metal ions. Asian J Chem 26(15):4901–4906. https://doi.org/10.14233/ajchem.2014.17024

Wang Y, Liu X, Wang H, Xia G, Huang W, Song R (2014) Microporous spongy chitosan monoliths doped with graphene oxide as highly effective adsorbent for methyl orange and copper nitrate ($Cu(NO_3)_2$) ions. J Colloid Interface Sci 416:243–251. https://doi.org/10.1016/j.jcis.2013.11.012

Wang H, Yuan X, Wu Y, Chen X, Leng L, Wang H, Li H, Zeng G (2015) Facile synthesis of polypyrrole decorated reduced graphene oxide–Fe_3O_4 magnetic composites and its application for the Cr(VI) removal. Chem Eng J 262:597–606. https://doi.org/10.1016/j.cej.2014.10.020

Wang H, Gao B, Fang J, Ok YS, Xue Y, Yang K, Cao X (2018) Engineered biochar derived from eggshell-treated biomass for removal of aqueous lead. Ecol Eng 121:124–129. https://doi.org/10.1016/j.ecoleng.2017.06.029

Wang Y, Li A, Cui C (2021) Remediation of heavy metal-contaminated soils by electrokinetic technology: Mechanisms and applicability. Chemosphere 265:129071. https://doi.org/10.1016/j.chemosphere.2020.129071

Wang J, Zhang J, Han L, Wang J, Zhu L, Zeng H (2021) Graphene-based materials for adsorptive removal of pollutants from water and underlying interaction mechanism. Adv Colloid Interface Sci 289:102360. https://doi.org/10.1016/j.cis.2021.102360

Wei M, Chai H, Cao Y, Jia D (2018) Sulfonated graphene oxide as an adsorbent for removal of Pb2+ and methylene blue. J Colloid Interface Sci 524:297–305. https://doi.org/10.1016/j.jcis.2018.03.094

Xu Y, Wu Q, Sun Y, Bai H, Shi G (2010) Three-dimensional self-assembly of graphene oxide and DNA into multifunctional hydrogels. ACS Nano 4(12):7358–7362. https://doi.org/10.1021/nn1027104

Xu J, Lv H, Yang S-T, Luo J (2013) Preparation of graphene adsorbents and their applications in water purification. Rev Inorg Chem 33(2–3):139–160. https://doi.org/10.1515/revic-2013-0007

Xu W, Wang J, Wang L, Sheng G, Liu J, Yu H, Huang X-J (2013) Enhanced arsenic removal from water by hierarchically porous CeO_2–ZrO_2 nanospheres: Role of surface- and structure-dependent properties. J Hazard Mater 260:498–507. https://doi.org/10.1016/j.jhazmat.2013.06.010

Yang S-T, Wang H, Guo L, Gao Y, Liu Y, Cao A (2008) Interaction of fullerenol with lysozyme investigated by experimental and computational approaches. Nanotechnology 19(39):395101. https://doi.org/10.1088/0957-4484/19/39/395101

Yang S-T, Chang Y, Wang H, Liu G, Chen S, Wang Y, Liu Y, Cao A (2010) Folding/aggregation of graphene oxide and its application in Cu2+ removal. J Colloid Interface Sci 351(1):122–127. https://doi.org/10.1016/j.jcis.2010.07.042

Yang X, Wan Y, Zheng Y, He F, Yu Z, Huang J, Wang H, Ok YS, Jiang Y, Gao B (2019) Surface functional groups of carbon-based adsorbents and their roles in the removal of heavy metals from aqueous solutions: A critical review. Chem Eng J 366:608–621. https://doi.org/10.1016/j.cej.2019.02.119

Yu J-G, Yu L-Y, Yang H, Liu Q, Chen X-H, Jiang X-Y, Chen X-Q, Jiao F-P (2015) Graphene nanosheets as novel adsorbents in adsorption, preconcentration and removal of gases, organic compounds and metal ions. Sci Total Environ 502:70–79. https://doi.org/10.1016/j.scitotenv.2014.08.077

Yu R, Shi Y, Yang D, Liu Y, Qu J, Yu Z-Z (2017) Graphene oxide/chitosan aerogel microspheres with honeycomb-cobweb and radially oriented microchannel structures for broad-spectrum and

rapid adsorption of water contaminants. ACS Appl Mater Interfaces 9(26):21809–21819. https://doi.org/10.1021/acsami.7b04655
Zhang L, Li X, Huang Y, Ma Y, Wan X, Chen Y (2010) Controlled synthesis of few-layered graphene sheets on a large scale using chemical exfoliation. Carbon NY 48(8):2367–2371. https://doi.org/10.1016/j.carbon.2010.02.035
Zhang Y, Liu Y, Wang X, Sun Z, Ma J, Wu T, Xing F, Gao J (2014) Porous graphene oxide/carboxymethyl cellulose monoliths, with high metal ion adsorption. Carbohydr Polym 101:392–400. https://doi.org/10.1016/j.carbpol.2013.09.066
Zhang L-H, Sun Q, Yang C, Lu A-H (2015) Synthesis of magnetic hollow carbon nanospheres with superior microporosity for efficient adsorption of hexavalent chromium ions. Sci China Mater 58(8):611–620. https://doi.org/10.1007/s40843-015-0076-8
Zhang C-Z, Yuan Y, Guo Z (2018) Experimental study on functional graphene oxide containing many primary amino groups fast-adsorbing heavy metal ions and adsorption mechanism. Sep Sci Technol 53(11):1666–1677. https://doi.org/10.1080/01496395.2018.1436071
Zhao D, Gao X, Wu C, Xie R, Feng S, Chen C (2016) Facile preparation of amino functionalized graphene oxide decorated with Fe3O4 nanoparticles for the adsorption of Cr(VI). Appl Surf Sci 384:1–9. https://doi.org/10.1016/j.apsusc.2016.05.022
Zhao S, Huang W, Xie J, Liu Z, Qu Z, Yan N (2021) Mercury removal from flue gas using UiO-66-type metal-organic frameworks grafted with organic functionalities. Fuel 289:119807. https://doi.org/10.1016/j.fuel.2020.119807

Chapter 3
Graphene Nanocomposites and Their Role in Heavy Metal Clean-Up

3.1 Graphene Nanocomposites

Composite material is a product of binding together two or more discrete materials of different states (chemically or physically) and exhibits superior properties as compared to their components. Nanocomposite refers to a composite material whose one or more constituent materials are in the nanometre scale. For instance, nanoparticles (NPs) with a size less than 100 nm are integrated with other 2 two-dimensional (2D) materials/substrates for developing nanohybrids. The developed nanocomposites are enriched with the physicochemical attributes, thereby highly suitable for applications in diverse fields, including environmental remediation. In a similar direction, graphene nanocomposites can be developed by surface engineering of the graphene framework with constituents like metallic or non-metallic NPs, ceramic materials, polymer matrix, etc. This technique is fundamental for embellishing the properties of pristine graphene for achieving desired outcomes in the field of heavy metal (HM) ions adsorption and removal in a sustainable manner (Ravi et al. 2020). Different types of graphene nanocomposites are classified as polymeric, metallic, organic, and inorganic composites of GO and rGO.

Polymer-based functionalized GO: owing to the splendid mechanical, electronic, and thermal attributes, graphene works as a prominent filler material for composite development. Gr/P nanocomposites' fabrication and application have increased tremendously over the last decade (Khatun et al. 2020; Ibrahim et al. 2021). Enhanced attributes of polymer composites with the addition of graphene are mentioned.

Mechanical properties: With the integration of graphene, the resin modulus of the composite material improved significantly. For instance, the modulus of epoxy resin was observed to be enhanced by nearly 31% with the integration of 0.1% (wt%) of graphene platelets (Rafiee et al. 2009; Shen et al. 2013). The superior mechanical attributes of the nanocomposite are due to the trapping of graphene in its structure (Adak et al. 2018). Wang and co-workers reported the significant impact of graphene

P. Mahajan and V. Khanna, *Graphene in the Environment*, Environmental Science and Engineering, https://doi.org/10.1007/978-3-031-96708-5_3

nanoplatelet size on the strength, flexural modulus, and tensile of a graphene/epoxy nanocomposite.

Thermal properties: It is observed that polymer composites are poor thermal conductors, which hinders their wide-scale applications in diverse fields. As graphene materials hold immense potential as thermal interface materials involved in heat dissipation, the integration of polymer composites with graphene NMS provides a great opportunity for preparing advanced functional materials (Atif et al. 2016). An increase of nearly two orders of magnitude (0.2–20 W/mK) in the thermal conductivity of the polymer has been noticed with integration of graphene at 30 wt%, provided in the form of a graphene film, inducing thermal stability to the composite (Bustero et al. 2020).

Electrical properties: Providing graphene as filler in the polymer matrix elevates the electrical conductivity of the composite owing to its superior conductivity and aspect ratio. Graphene inculcates a conductive network in the polymer matrix. One of the studies reported that the electrical conductivity of carbon/epoxy nanocomposite reached 0.70 S/cm with the addition of 5 mg/ml graphene (Li et al. 2019).

A research group prepared GO/polyaniline (GO/PANI) nanocomposite via chemical technique and conducted its characterisation via FESEM, XRD, FTIR, and TGA methods. This nanocomposite worked as an excellent adsorbent in the presence of a chelating agent, viz., dithizone ligand, for quantification of Cd(II) ions from the aqueous media (Farajvand et al. 2023). Besides, HM ion sensors were also fabricated using novel hydrogels made up of rGO and conjugate polymer polypyrrole (PPy) composite and applied for the detection of four different HM ions, such as Pb, Cu, Cd, and Hg (Lu et al. 2018). To prevent the aggregation of GO nanosheets, a GO nanocomposite was developed using poly(3-aminopropyltriethoxysilane) (PAS) oligomers by cross-linking them with GO nanosheets in a 3d framework. This generated highly efficient adsorbent, viz., PAS-GO. A dimensional crosslinking framework was created to avoid aggregation of GO nanosheets, which facilitates its facile interaction with external molecules as well as inserting a large number of amino functional groups for adsorbing Pb(II) ions with great efficiency (Luo et al. 2014).

Organic molecule-based functionalized GO: The functionalization of graphene with organic molecules is carried out via non-covalent or covalent modifications. One of the most commonly used organic molecules for developing GO nanocomposites is ethylenediaminetetraacetic acid (EDTA). It provides GO with innumerable surface functional groups, facilitating elevated HM ions adsorption. Besides, cetyltrimethylammonium bromide (CTAB) is capable of prohibiting the agglomeration of GO, thereby advancing the process of adsorption smoothly. Various other organic molecules, such as aminopyrazole, triisopropanolamine, etc., also play a significant role in developing GO composites, thereby elevating their adsorption capacity (Krishna et al. 2023).

Inorganic molecule-based functionalized GO: Different inorganic materials, such as ceramics, metals, and metal oxides, are involved in the production of graphene nanocomposites with magnificent adsorption abilities. Amongst these, magnetic inorganic components/materials are preferred extensively owing to their facile separation or regeneration after completion of the adsorption process. It is observed that inorganic NMS with excellent magnetic attributes, as well as catalytic capability, are considered and applied more frequently for developing GO composites with excellent adsorption ability. For instance, iron oxide-functionalized GO and rGO have reportedly been used for the efficient and sustainable removal of HM ions. Besides, oxides of silicon, titanium, manganese, layered double hydroxides (LDHS), and various semi-conducting materials such as cadmium sulphide also work as wonderful materials for the functionalization of graphene (Krishna et al. 2023).

3.2 Synthesis Techniques for Graphene Nanocomposites

In situ polymerisation: this nanocomposite synthesis technique exhibits great potential, especially in developing heterogenous graphene in a polymer framework (Fig. 3.1a). This involves polymerisation of monomers in the presence of a stacking material. For this process, the monomer solution is fed with filler, causing its expansion. As there is swelling in fillers, monomers occupy the spaces in between their layers. For even distribution, the solution is sonicated and stirred, followed by initiation of the polymerisation process via radiation, heat, initiator diffusion, and catalyst.

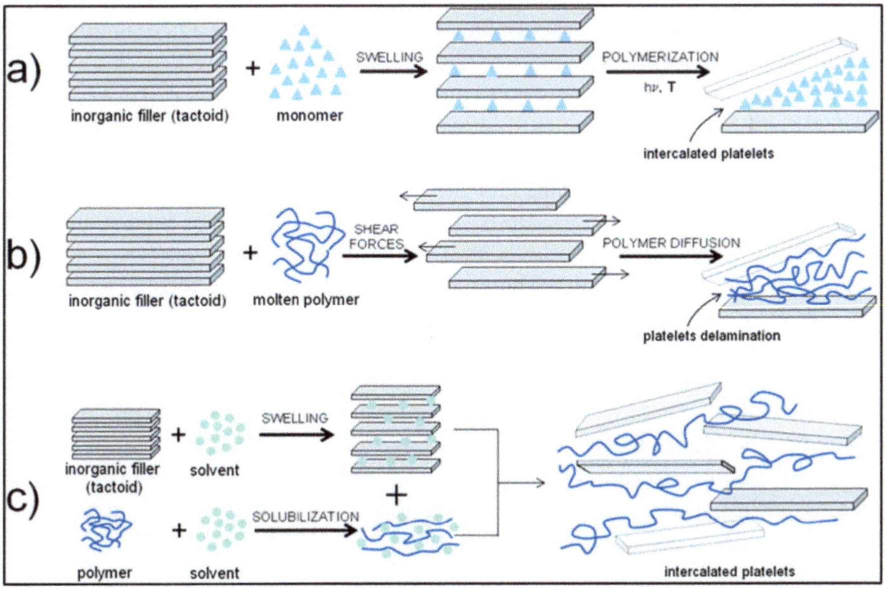

Fig. 3.1 Pictorial view of different synthesis mechanisms for nanocomposites such as **a** in situ polymerisation, **b** melt processing, and **c** exfoliation adsorption (Ibrahim et al. 2021; Uysal Unalan et al. 2014)

Thereafter, polymer chains are developed around and in between filler layers; however, they keep the layers apart for better exfoliation. Besides, the introduction of additional functional groups during polymerisation confers outstanding compatibility and dispersion between graphene and polymer. The advantage of this process as compared to previous ones is the usage of thermoset polymers as well as maintaining robust covalent bonding amongst the polymer matrix and nanofiller (Uysal Unalan et al. 2014; Fu et al. 2019).

Melt blending/intercalation: This process is suitable for developing polymer hybrids possessing inorganic nanoparticles (NPs) (Avouris and Xia 2012). This is a conventional technique which is economic and eco-friendly as it does not involve solvent usage. Hence, graphene or its derivatives can be immediately integrated with the molten polymer matrix. The major drawback of this technique lies in the graphene's poor dispersion in the polymer matrix, particularly in the case of elevated filler loadings, leading to extreme viscosity of composites. Melt blending is the most frequently applied method for the preparation of polymer nanocomposites, which involves the generation of a viscous liquid through the melting of polymer (Paul and Robeson 2008). Afterwards, using a high shear mixer as well as elevated temperature, the dispersion of nanofillers is conducted in the polymer matrix (Fig. 3.1b) (Yuan et al. 2009). High temperature-resistant or stable fillers are preferred to avoid their degradation during the polymerisation process. A researcher group prepared polylactic acid (PLA)/graphene nanocomposite through this process, and it exhibited superior mechanical attributes, electrical conductivity, and crystalline structure (Bao et al. 2012). The major challenge with the melt blending process is extreme viscosity with larger filler concentration and a reduced rate of intercalation when compared to other techniques.

Exfoliation adsorption: for developing ultra-adherent forces in between filler materials and polymer matrix, exfoliation adsorption is a very effective technique (Fig. 3.1c) (Uysal Unalan et al. 2014; Vazquez et al. 2012). This process also involve the intermixing of filler materials and the matrix in the presence of a compatible solvent (Alexandre and Dubois 2000). The solvent facilitates the expansion of the filler, thereby widening the interlayer area and allowing the intercalation of polymer chains between the filler layers. The mixture thus obtained is sonicated and stirred with simultaneous removal of solvent via evaporation. Subsequently, the polymer chains get entrapped in the layers of filler, thereby creating a multilayered framework. Though this technique is suitable for developing nanocomposites with polymers exhibiting low polarity, it is highly unsuitable for industrial-scale production of nanocomposites due to the involvement of a large amount of solvents (Uhl and Wilkie 2002). Exfoliation adsorption technique can be applied for the synthesis of polymer nanocomposites with polymers viz., polyurethane (PU), poly(methylmethacrylate) (PMMA), and poly(vinyl alcohol) (PVA) (Kim et al. 2010; Zhang et al. 2012; Zhao et al. 2010).

Emulsion-based techniques: As compared to the above-mentioned techniques, the emulsion technique for nanocomposite preparation allows more uniform dispersion of the filler. Besides, it augments extremely intense interaction between the polymeric material and the filler, developing nanocomposites with greater mechanical strength and the least percolation threshold in comparison to aforementioned techniques at a similar rate of filler loading (Sengupta et al. 2011). Abundant oxygen-containing surface functional groups of GO, besides enhancing its dispersibility, also confers stabilisation to the polymer particles in the emulsion by acting as a surfactant (Fadil et al. 2022). The other attractive feature owed by emulsion technique involves application for wide range of polymers, accelerated polymerization rate as no hinderance is posed by the rate of filler content, the obtained nanocomposite matrix can be implied for attaining wide-area coverage on diverse substrate, and eco-friendly due to involvement of water rather than solvent, thus highly suitable for commercial level as well (Fig. 3.2).

Miniemulsion polymerisation: polymerisation is carried by monomer droplet nucleation, referring to the direct conversion of monomer droplets to polymer particles in a typical 50–500 nm diameter size range. This process is carried out using shear apparatus such as ultrasonifier, high pressure homogeniser, etc., which conducts the deterioration of oil phase into sub-micron oil droplets called miniemulsions. These droplets are stabilised by integrating them with surfactants, and they also possess co-stabilisers such as hexadecane for circumventing diffusional deterioration termed as Ostwald ripening (Taylor 1998). The excitement this process holds is the absence of monomers or any other reagents' diffusion across the aqueous phase, hence it holds immense potential to synthesise organic/inorganic hybrid materials/particles (Fadil et al. 2019; Cai et al. 2019).

Microemulsion polymerisation: This technique exhibits better thermodynamic stability as compared to miniemulsion polymerisation. Besides, the surfactant

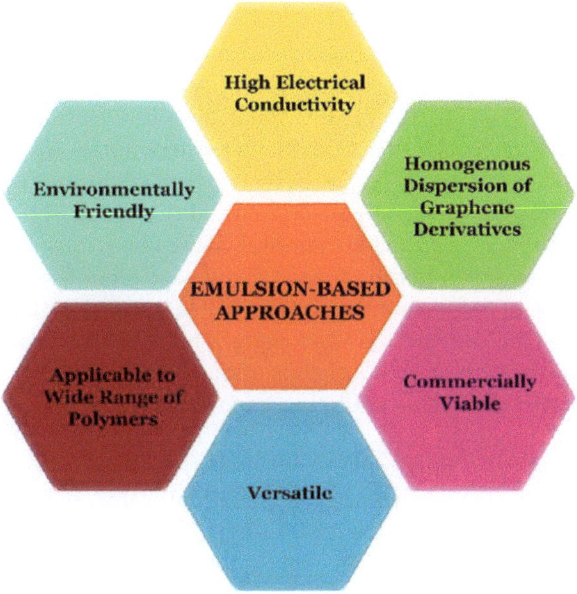

Fig. 3.2 Pictorial view of benefits of emulsion-based techniques in comparison with other polymer/graphene-based nanocomposite synthesis techniques (Fadil et al. 2022)

concentration as compared to the latter one is also higher, and it varies between 100 and 300 wt% corresponding to monomer. The polymerisation process is initiated via engulfing of monomers by non-nucleated monomer-swollen micelles through aqueous phase-mediated diffusion to enlarge polymer particles (Wakamatsu et al. 2007; Zetterlund et al. 2009). Further, the amalgamation process, to a limited extent, yields polymer particles of size 20–60 nm. In this regard, Patole and co-workers developed graphene sheets functionalized with polystyrene NPs via Microemulsion polymerisation involving the usage of pentanol, sodium dodecyl sulphate (SDS), and azobis(isobutyronitrile) (AIBN) working as surfactant and co-surfactant. The obtained graphene/polystyrene nanocomposite was later used as filler in introducing polystyrene matrix for developing composite films through solution casting (Patole et al. 2010).

Emulsion polymerisation: it is one of the most effortless and vigorous processes for developing polymer NPs on a wide scale. The initiators for this technique are soluble in the water or continuous phase, however, both monomer and polymer are insoluble. The process of polymer particle formation is categorised into micellar nucleation, which involves the use of surfactant, and homogenous nucleation, which does not involve any stabiliser or surfactant (Fadil et al. 2022).

3.3 Graphene-Based Nanocomposites for Heavy Metal Adsorption

3.3.1 Lead (Pb) Ions Removal

Graphene oxide/Fe_3o_4/chitosan nanocomposite (GFC): as shown in Fig. 3.3a, GFC nanocomposite has been developed successfully, and no Fe_3o_4 particles can be seen from TEM images of GFC (Vo et al. 2024). Besides, a synergistic role is played by chitosan and GO-Fe_3O_4 NPs; chitosan prohibits any kind of agglomeration of Fe_3o_4 NPs, whereas Fe_3o_4 checks on any reattachment of GO sheets. This nanocomposite exhibited outstanding adsorption ability towards Pb(II) ions from the aqueous media. There was nearly a 90% removal rate of Pb(II) ions with 30 mg/L GFC initial concentration at equilibrium time of 50 min and ph of 5, and an adsorbent concentration of 1 g/L at room temperature (27°C). The optimal Pb(II) ion concentration adsorbed by the GFC was nearly 63.45 mg/g. The reaction kinetics indicates it to be a second-order reaction, thus a chemisorption process, and thermodynamic studies confirm a higher adsorption rate at low temperature, particularly room temperature. Moreover, the Langmuir model fits best in comparison to the Freundlich isotherm model for the process of adsorption, which was reported to be an exothermic reaction as indicated by a negative $\Delta H°$ (22.12 kj/mol) value. In addition, after four adsorption and desorption cycles, the Pb(II) ions removal rate by GFC was found to decline only by 12%, hence implies its great stability and regeneration, suitable for industrial scale application (Vo et al. 2024).

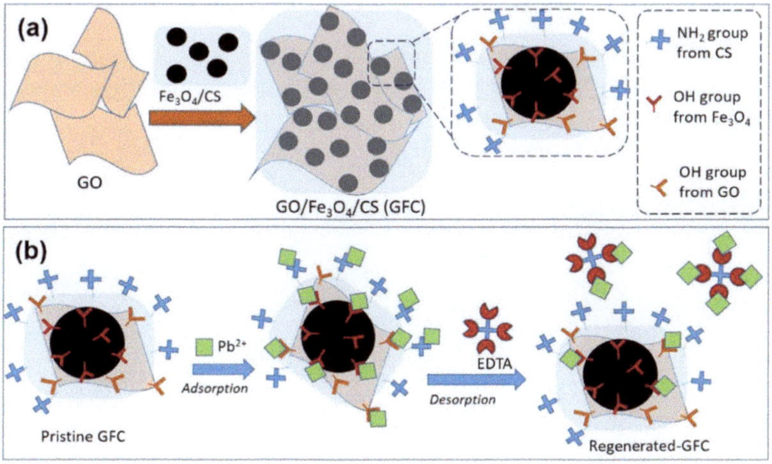

Fig. 3.3 **a** Illustrative representation of GFC formation, its structure and specific function groups and **b** Techniques for its application as a smart and regenerable nanosorbent, particularly for Pb(II) adsorption (Vo et al. 2024).

Functionalized metal organic framework and magnetic graphene oxide (Fe_3O_4@C-GO-MOF) nanocomposite: the schematic preparation of Fe_3O_4@C-GO-MOF nanocomposite is illustrated in the Fig. 3.4.

From Fig. 3.5a, the smooth deposition of Fe_3O_4@C NPs can be seen on the GO layered sheet's crinkled surface. The transmission electron micrograph of Fe_3O_4@C-GO (Fig. 3.5d) further highlights spherical Fe_3O_4@C NP's embellished on GO crinkled sheets randomly. The typical Uio66-NH_2 cubic structure is also obtained (Fig. 3.5b). The TEM image of the UiO66-NH_2 NPs is visible in Fig. 3.5e. The scanning electron micrograph of Fe_3O_4@C-GO-MOF nanocomposite is shown in Fig. 3.5c, wherein Uio66-NH_2 crystals are seen to be aggregated with GO sheet possessing Fe_3O_4@C NPs, as evidenced by earlier studies (Cao et al. 2018; Lin et al. 2018). Figure 3.5f displays attached Uio66-NH_2 particles onto the transparent GO

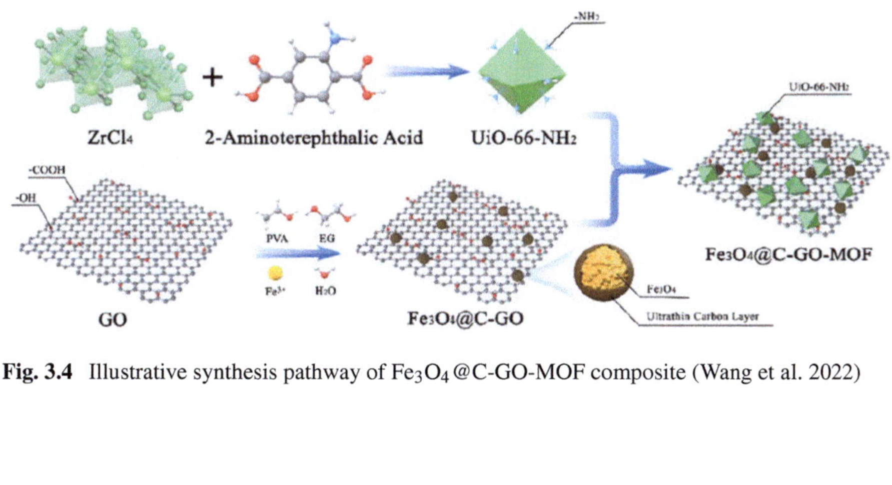

Fig. 3.4 Illustrative synthesis pathway of Fe_3O_4@C-GO-MOF composite (Wang et al. 2022)

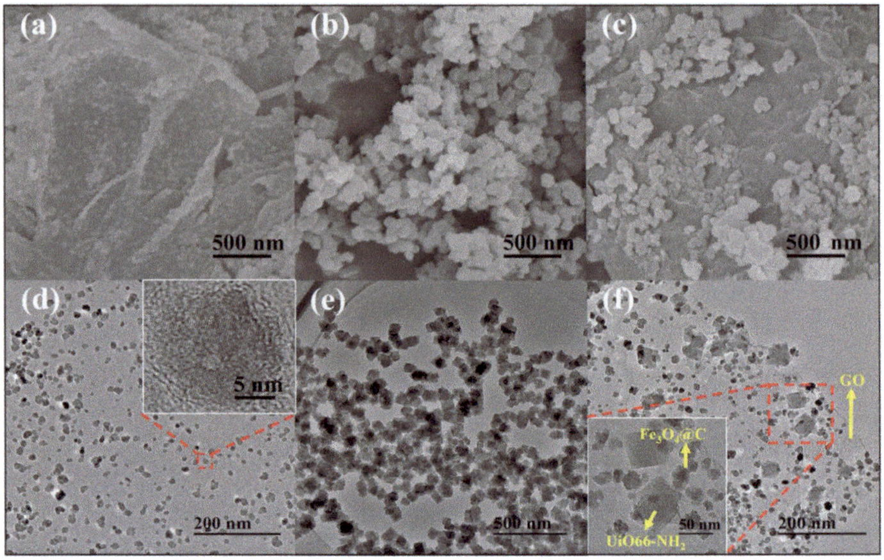

Fig. 3.5 Represents scanning electron micrographs of **a** Fe_3O_4@C-GO, **b** Uio66-NH_2, and **c** Fe_3O_4@C-GO-MOF composite; transmission electron micrographs of **d** Fe_3O_4@C-GO, **e** Uio66-NH_2, and **f** Fe_3O_4@C-GO-MOF composite (Wang et al. 2022)

lamellar structure [inset Fig. 3.5f], evidencing characteristic structural integrity of these particles (Wang et al. 2022).

Amino (UiO66-NH_2 NPs) and magnetic GO (Fe_3O_4@C-GO-MOF) functionalized novel nanocomposite was synthesised by a group of researchers for analysing the efficient removal of Pb ions. The obtained nanocomposite structure displayed a large surface area of nearly 332.260 m^2/g and high porosity of nearly 0.271 cm^3/g. The optimal adsorption ability by Fe_3O_4@C-GO-MOF composite was exhibited at ph 6, 10 mg concentration, and a contact time of nearly 1 h. By analysing the adsorption isotherm, it was predicted that the adsorption ability is a function of the adsorption sites on the nanocomposite surface. In the present scenario, the adsorption intensification is governed by the oxygen-containing functional groups of GO and amino-containing groups of Uio66-NH_2 and the maximal Pb ions adsorption rate was estimated to be 344.83 mg/g. The presence of magnetic NPs also facilitates the facile separation of this composite from the solution. Besides, Fe_3O_4@C-GO-MOF possessed tremendous selectivity towards Pb ions, which were adsorbed amongst seven other co-existing ions, and excellent recovery output, even up to five adsorption–desorption cycles (Wang et al. 2022).

3.3.2 Arsenic (As) Ions Removal

Graphene oxide functionalized Uio-66-NDC (Uio-66-NDC/GO) nanocomposites: Singh and co-workers developed a novel nanocomposite comprising GO and zirconium (Zr)-based MOF (Uio-66-NDC) for analysing its ability towards As(V) adsorption (Singh et al. 2022).

Although the developed nanocomposite exhibited good adsorption tendency across a large ph range, i.e., from highly acidic 1 to basic 10, the maximal output was found at ph 3. The As(V) adsorption concentration obtained for Uio-66-NDC/GO was found to be 147.06 mg/g, much higher than the earlier reported commercial adsorbents. The adsorption kinetics of the As(V) adsorption on Uio-66-NDC exhibited a pseudo-second-order model, indicating a chemisorption mechanism. The positive standard enthalpy (ΔH°) indicates the endothermic nature of As(V) adsorption mechanism on the Uio-66-NDC nanocomposite.

After five adsorption–desorption cycles, the removal efficiency of this nanocomposite for As(V) ions remained only around 52.24% as compared to the starting efficiency, evidencing its exorbitant regeneration tendency (Fig. 3.6) (Singh et al. 2022).

Graphene oxide-Iron nanohybrid (GFeN): The synthesis of GFeN was carried out via a sol–gel mechanism, which involved embellishing the GO surface with iron and iron oxide NPs. Through kinetic studies, it was revealed that within the first 6 h and 8 h, nearly 99% removal of As(III) and As(V), respectively, was noticed with an initial As concentration of 5 mg/L (Fig. 3.7a). Three distinct phases were involved for adsorption of both As species viz., (i) quick removal of As ions in first hour exhibiting more than 50% removal rate (Fig. 3.7a inset); (ii) comparatively slow adsorption rate after 1 h up to 6–8 h; (iii) extremely minute (insignificant) adsorption rate beyond 6–8 h. Moreover, adsorption isotherms used for understanding the adsorption

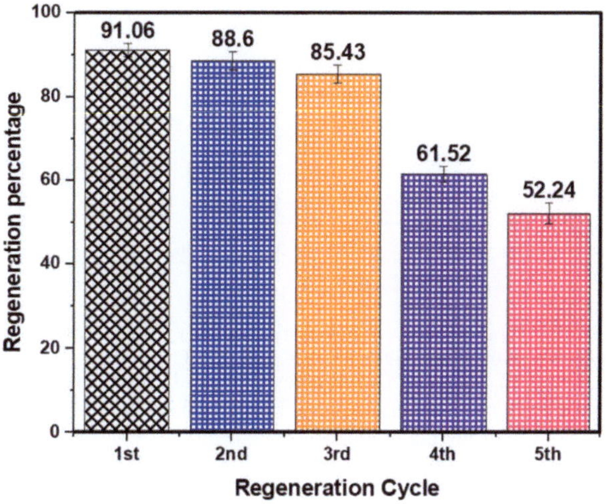

Fig. 3.6 Bar graph representing the recycling and reusability of Uio-66-NDC/GO for As(V) removal (Singh et al. 2022)

mechanism of As by GFeN revealed that, as compared to the Freundlich isotherm model, the adsorption data for both the As species fit better for the Langmuir model, evident via R^2 value (Fig. 3.7b and c). Isotherm data revealed the adsorption ability of GFeN ranged nearly 306.10 and 431.41, respectively, for As(III) and As(V) (Das et al. 2020).

The predicted mechanism, which exists in the adsorption As through GFeN, is illustrated in Fig. 3.8 (Das et al. 2020). The reduction potential in the case of As(III) to As(0) is 0.499 V, and it is -0.477 V in the case of Fe(0) to Fe(II), favouring the earlier reaction as compared to later as it is favoured thermodynamically (Melitas et al. 2002; Sasaki et al. 2009). A mixture of crystalline and amorphous phases of iron oxides exists in the oxide layer of the GFeN, the amorphous phase being the dominant one, nearly 2 to 10 times more than the other phase (Ling and Zhang 2014; Wu et al. 2017). There is rapid ionic movement of As species induced by the porosity amorphous phase and oxygen spaces in the case of the crystalline phase (Kerisit and Rosso 2005). Thereafter, both As species are reduced to an intermetallic point of

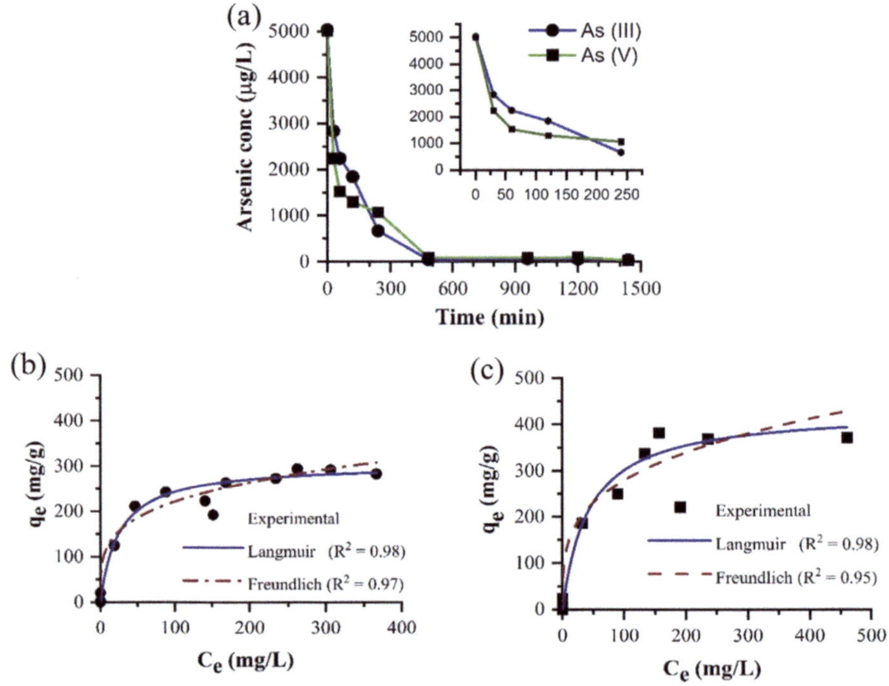

Fig. 3.7 Graphical representation of **a** Arsenic removal by GFeN over time at an initial As concentration (C0) of 5 mg/L (Inset: gives magnified data of first 2 h), **b** Adsorption isotherms for As(III) and **c** As(V) using Langmuir and Freundlich isotherms. The Langmuir model provided a better fit for both, yielding maximum adsorption capacities of 306.10 ± 9.92 mg/g for As(III) and 431.41 ± 25.95 mg/g for As(V). Experimental conditions: Adsorbent dose 1/4 250 mg/L and Initial ph 1/4 7 (Das et al. 2020)

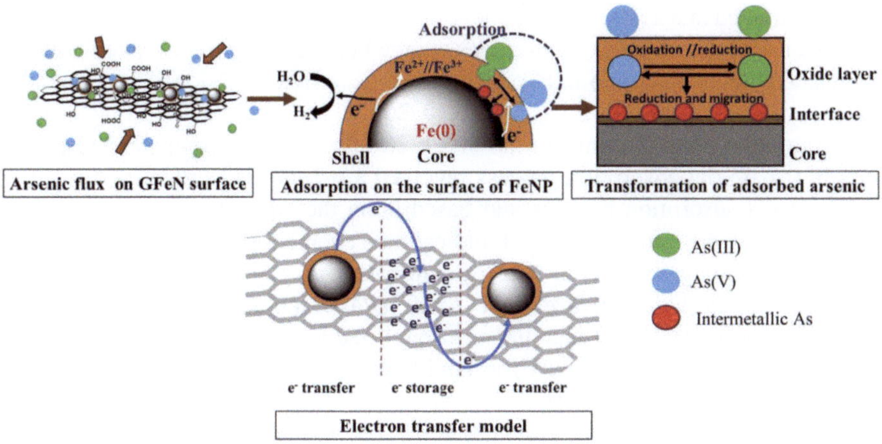

Fig. 3.8 Top: represents the proposed underlying mechanism for arsenic (As) removal by GFeN. The involves As accumulation at the solution-adsorbent interface, thereafter its adsorption onto the FeNPs embedded on GFeN surface. Concurrently oxidation and reduction occurs of adsorbed As, and migration of the reduced As happens into FeNPs core. Bottom: gives overview of graphene oxide (GO) sheet serving as electron reservoir liberated during Fe oxidation, and the utilization of the stored electrons is done by the oxide layer (on FeNPs) to replenish itself for further adsorption of As (Das et al. 2020)

As(0) via their transfer towards the Fe^{o} core. The obtained intermetallic phase of As species exhibits great stability as compared to other species, adding to the adsorption efficiency of GFeN (Yan et al. 2012; Tuček et al. 2017).

3.3.3 Chromium (Cr) Ions Removal

Magnetic-graphene oxide nanocomposite (GO-Fe_3o_4): A scientific group prepared GO-Fe_3o_4 nanocomposite from *Schleichera oleosa* (kesambi wood) acquired from Timor Island, Indonesia (Neolaka et al. 2020). This wood provided raw material for synthesising graphite, which later transformed to GO applying the Hummers method (Sheikhmohammadi et al. 2017; Zhang et al. 2016). The magnetic composite was then prepared by the technique of co-precipitation (Jaleh et al. 2017; Sammaiah et al. 2018). The adsorption strategy for Cr(VI) ions onto the GO-Fe_3o_4 surface is governed by electrostatic interactions mediated by positively charged active spots, which elevates the tendency for adsorbing Cr(VI) ions owing to protonated functional groups. The major complexation of Cr(VI) ions occurs with oxygen-containing functional groups and facilitates electrostatic attraction with iron oxide (Fe_3o_4). In the present scenario, the Cr(VI) adsorption is carried by electrostatic anion-cation attraction (Yang et al. 2020). In the first instance, electrostatic anion-cation attraction leads to Cr(VI) adsorption followed by its reduction to Cr(III) ions which were adsorbed at the last instance by an adsorbent (Yang et al. 2019).

It is estimated that acidic groups, phenolic groups, and an elaborate system of GO-Fe_3o_4 facilitate the process of anionic exchange between the proton linked with the oxygenated functional group and Cr(VI) species, as well as reduction of hexavalent Cr ions into trivalent form (Fig. 3.9) (Zhou et al. 2017). The maximal Cr(VI) adsorption rate occurred at ph of 2, contact time of 1 h 20 min, and nanocomposite adsorbent mass of 0.1 g. The adsorption process involved a pseudo-first-order kinetic model estimated via adsorption kinetics and based upon the adsorption isotherm, it was found to follow the Langmuir model, indicating a chemisorption mechanism for the Cr(VI) adsorption process by GO-Fe_3o_4 nanocomposite (Neolaka et al. 2020).

Porous polyacrylonitrile/graphene oxide (PAN/GO) nanofibers: Feng et al. developed porous PAN/GO nanofibers through facile electrospinning of PAN/GO sheets solution in N, N-dimethylformamide (DMF)/water (H_2O) mixed solvent (Fig. 3.10) (Feng et al. 2020). The synthesised nanocomposite nanofibers possessed a significantly high Brunauer–Emmett–Teller (BET) surface area in the range of nearly 164.2 $m^2\ g^{-1}$ as compared to only 11.7 $m^2\ g^{-1}$ of PAN nanofibers. Besides, the reusability capacity of this nanocomposite is outstanding, which is around 20 times.

Scanning electro-micrographs as shown in Fig. 3.11, display the surface morphologies of diverse nanofibers. The PAN nanofibers possessed a smooth surface and regular diameter distribution with a mean diameter of nearly 475 nm (Fig. 3.11a). However, the addition of water in the polymer solution (nearly 8 wt%, mass of

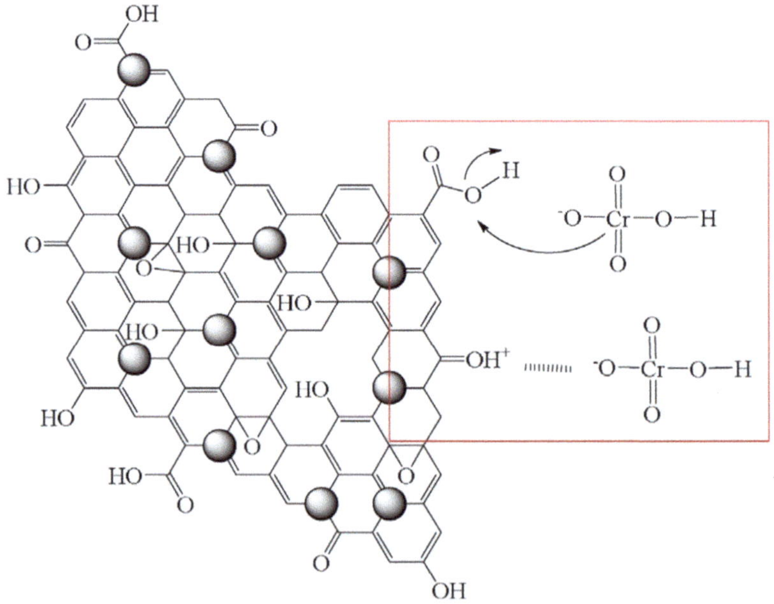

Fig. 3.9 Illustrative view of interaction occurring between Cr(VI) and GO-Fe_3O_4 (Neolaka et al. 2020)

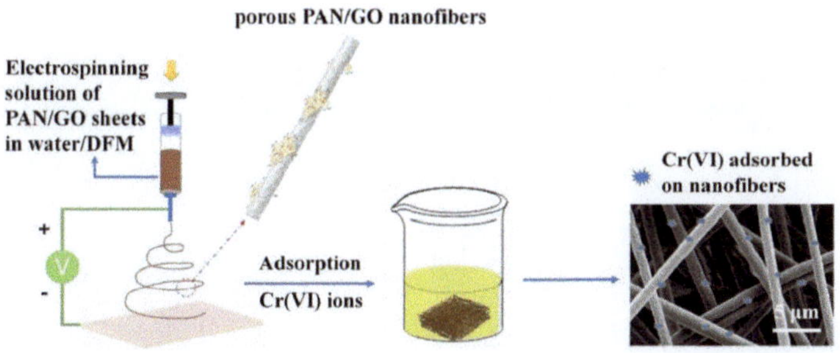

Fig. 3.10 Diagrammatic view of electrospinning process for obtaining porous PAN/GO nanofibers and its adsorption tendency for Cr(VI) ions in an aqueous media (Feng et al. 2020)

water to DMF) induced nanopores (nearly 10 nm in size) on the nanofiber surface, which also exhibited a rough surface and enlarged diameter, nearly 1356 nm (Fig. 3.11b). Further, integration of GO sheets (1.98 wt%, mass of GO to PAN) to the electrospinning solution induces abundant nanopores on the porous PAN/GO nanofibers (Fig. 3.11c–c3) with elevated size dissemination ranging from 5 to 20 nm (Fig. 3.11c1) with reduced diameter of the porous PAN/GO nanofibers (Fig. 3.11c2). Moreover, SEM and TEM images of these synthesised nanocomposite nanofibers exhibited eccentric porous structure amid nanofibers with a size ranging between 10–200 nm (Fig. 3.11c3) (Feng et al. 2020).

The Cr(VI) ions adsorption ability of porous PAN/GO nanofibers reached up to 382.5 mg/g, the topmost experimental value exhibited by functional nanofiber adsorbents. The adsorption reaction was quick and endothermic proceeded in a multi-molecular layer, indicating the involvement of the Freundlich model. Besides, the high reusability of this nanofiber was analysed via retention of more than 75% desorption capacity for Cr(VI) ions even after 20 adsorption–desorption cycles. This is an eco-friendly technique for environmental remediation as no harsh and toxic chemicals have been utilised in the production of porous PAN/GO nanofibers, thus avoiding any secondary pollution (Feng et al. 2020).

3.3.4 Mercury (Hg) Ions Removal

Thiourea-functionalized graphene oxide (GO-SH) nanocomposite: A scientific group prepared thiourea-functionalized GO (G3DTF) using thiourea-formaldehyde polymer (Witomska et al. 2019). As this nanocomposite possesses abundant thiol and amine functional groups attached to the surface of GO in addition to a large specific surface area, it exhibits great potential in HM ions adsorption. Goncalves et al. utilised this nanocomposite for the removal of Hg ions from real water samples such as sea water, bottled water, and ultrapure water. It was observed that G3DTF at 10 mg/L concentration were able to adsorb nearly 99% of Hg ions with an initial

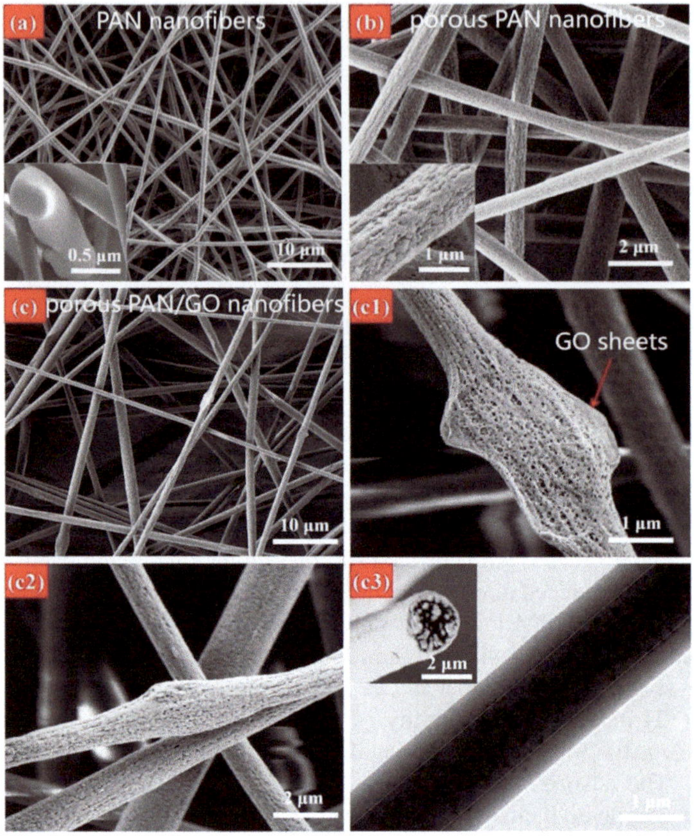

Fig. 3.11 Represents the scanning and transmission electron micrographs of **a** PAN, **b** porous PAN, **c** porous PAN/GO nanofibers; (c1–3) depicts the porous framework on the shell and core of the porous PAN/GO nanofibers, inset: cross-section (Feng et al. 2020)

content of 50 μg/L in all three real water samples, indicating its magnificent role in environmental remediation (Gonçalves et al. 2024).

Thiol-functionalized graphene oxide nanocomposite: owing to hard-soft acid base theory, there exists a strong attraction between sulphur-based functional groups, viz., sulfonic acid, sulfhydryl, etc. and Hg ions; hence, preparing adsorbent material with abundant sulphur-containing functional groups can impart excellent Hg ions removal ability. In this regard, Sun and co-workers synthesised thiol-functionalized GO (GO-SH) nanocomposite and studied its efficacy in adsorption of Hg ions from the aqueous media. With the initial Hg concentration of 10 mg/L, the adsorption ability of GO-SH was nearly 98.7% within 36 h, significantly high as compared to 52.8% in the case of GO alone (Fig. 3.12a). Further, in Fig. 3.12b, the adsorption curve's pattern for both GO and GO-SH is same in which the whole process of adsorption involves initial fast adsorption in which Hg ions are rapidly adsorbed by

GO and GO-SH which lasts for around 4 h. Whereas, late slow equilibrium process involves decline in the adsorption process which ultimately enters the equilibrium phase. It is observed that the Hg(II) adsorption tendency at equilibrium was nearly 26.48 mg/g in the case of GO, and 49.68 mg/g in the case of GO-SH, indicating much better adsorption capacity of the latter as compared to the former (Sun et al. 2023).

The adsorption process of Hg(II) ions on the GO-SH nanocomposite follows the Freundlich isotherm model and the pseudo-second order kinetic model. Besides, the occurrence of three linear segments viz., Hg(II) ions movement onto the GO-SH surface from the solution which also involves chemical adsorption, the second stage represented by the slow adsorption in which intraparticle diffusion occurs maintain the adsorption rate, and in the last stage i.e., final equilibrium stage occurs the slow diffusion of Hg ions from larger pores to smaller pores (Fig. 3.12c) (Sun et al. 2023).

3.3.5 Copper (Cu) Ions Removal

Magnetic chitosan/graphene oxide nanocomposites (MCGON): for preparing MCGON, Hosseinzadeh and Ramin first of all introduced magnetic iron oxide (Fe_3O_4) NPs to GO, which gives an advantage of effortless solid–liquid phase segregation. However, this process decreases the adsorption tendency of GO due to the occupancy of adsorptive sites by magnetic NPs (Hosseinzadeh and Ramin 2018; Parida et al. 2012; Wang et al. 2014). Hence, modifications in the GO structure using ethylenediamine and thereafter co-polymerisation with free radical, viz., acrylic monomers onto chitosan backbones, facilitate the high adsorption tendency of GO along with convenient segregation of magnetic NPs, lowering GO dispersibility in the aqueous medium and protecting the environment from its toxic impacts. Owing to enlarged specific surface area (132.9 $m^2\ g^{-1}$), superb pore volume (4.03 $cm^3\ g^{-1}$) and minute particle dimension (15 nm), MCGON exhibits extremely high adsorption capacity in the range 217.4 mg g^{-1} for Cu (II) ions. Besides, the magnificent superparamagnetic attribute of MCGON adsorbent is suitable for its separation with the help of an external magnetic source. The adsorption output could be ascribed to the pseudo-second order model, as well as followed Langmuir isotherm model. The adsorption of Cu (II) exhibited direct dependence on the temperature, indicated by enhanced adsorption with increasing temperature. The overall kinetics of the adsorption process indicates the Cu adsorption onto the MCGON adsorbent to be a prompt, endothermic, and physical phenomenon (Hosseinzadeh and Ramin 2018).

3.3.6 Cadmium (Cd) Ions Removal

Magnetic iron oxide/graphene oxide (Fe_3O_4/GO) nanocomposites: a nanocomposite, viz., Fe_3O_4/GO, was developed by a group of researchers for combining

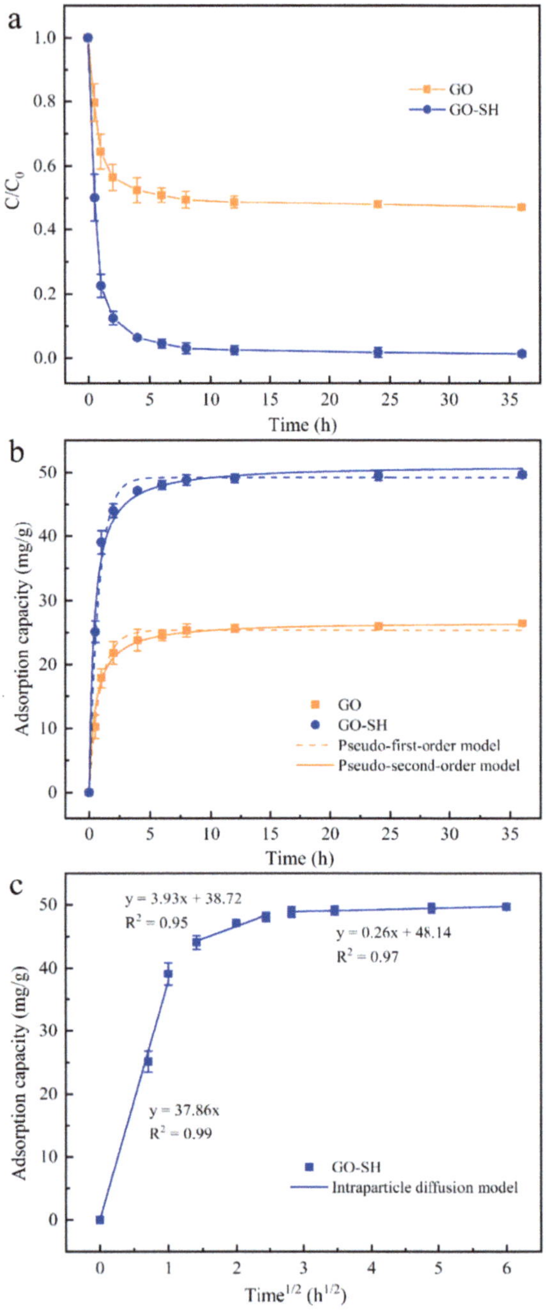

Fig. 3.12 Graphical representation of **a** Hg(II) ions removal efficiency of GO and GO-SH, **b** kinetic model fitting curves for Hg(II) removal by GO and GO-SH, and **c** intraparticle diffusion model analysis for GO-SH (Sun et al. 2023)

the superparamagnetic attributes of Fe_3O_4 and the high adsorption capacity of GO for adsorption of Cd ions. This nanocomposite can be segregated with an external magnetic source with ease (9). The co-precipitation process was involved for developing Fe_3O_4/GO nanocomposite via insertion of Fe_3O_4 NPs on the surface of GO sheets. The most suitable ratio of Fe_3O_4 to GO was found to be 4:1, which was synthesised efficiently. The adsorption of Cd ions on this nanocomposite followed the Langmuir isotherm and pseudo-second order models, and based upon this, the adsorption was found to be in the range of 52.083 mg/g noticed at ph 8, much higher than the individual components of the nanocomposite. The Fe_3O_4 NPs exhibited a size range of nearly 10 to 15 nm, which were present on the GO surface obtained via different characterisation techniques. Moreover, the nanocomposite possessed a high recovery tendency from the aqueous medium (Thy et al. 2019) (Table 3.1).

In general, the mechanism involved for adsorption of HM ions by graphene nanocomposite involves almost similar forces and interactions as pristine graphene and its derivatives, mentioned in the previous chapter (Fig. 3.13).

3.4 Advantages of Graphene-Based Nanocomposites for Heavy Metal Ion Adsorption

Graphene-based nanocomposites exhibit outstanding adsorption abilities as compared to graphene and its derivatives (GO and rGO) (Fig. 3.14). It is observed that graphene-based polymer composites possess great mechanical strength, a highly porous framework, and exorbitant thermal stability. Besides, the lucrative hydrophilicity of these composites plays a pivotal role in the removal of HM ions through the process of complexation. For instance, GO-CNF aerogel composite possesses relatively high surface area as compared to its components, as well as prevents the restacking of GO sheets in composite form (Nie et al. 2021). High adsorption ability, although it marks the material to be a superior adsorbent, yet balancing other parameters is also important. A group of researchers reported CGO fibre composite to be a better adsorbent than GO, besides having less adsorption ability, however excellent stability and facile separation define CGO fibre composite as a superior adsorbent compared to GO (Chen et al. 2016). Graphene-based nanocomposites have overcome the problem of stability to a larger extent as compared to pristine graphene adsorbents as well as other conventional adsorbents.

The studies from the literature indicated that introduction of graphene derivatives to the polymers, particularly biopolymers such as chitin, cellulose, has tremendously enhanced the stability and strength of the obtained composite in addition to their marvellous adsorption capabilities (Han et al. 2019). Further, graphene composites possess high significance in rapid kinetics. For instance, a study reported that GO in combination with PANI, i.e., GO/PANI composite as an adsorbent, took around 20 min to achieve the adsorption equilibrium for Pb(II) ions and GO for the same took around 90 min (Yang et al. 2018). As recyclability and regeneration are also important

Table 3.1 Adsorption tendencies of various graphene-based materials for different heavy metal ions (Xu et al. 2022; Yang et al. 2019; Ayala-Claveria et al. 2024)

Material	Heavy metal ions	Adsorption tendency (mg/g)	References
GO	Cd(II)	35.7	Wang and Chen (2015)
GO	Cd(II)	44.64	Xue et al. (2016)
Magnetic chitosan/GO	Pb(II)	76.94	Fan et al. (2013)
Chitosan–functionalized GO	As(V)/As(III)	71.9 / 64.2	Kumar and Jiang (2016)
Magnetic GO (MGO)@APhen	Pb(II)	Removal efficiency of 97.2%	Abaszadeh et al. (2022)
MGO ion–imprinted polymer	As(III)	49.42	Chi et al. (2021)
MGO–EDTA	Pb(II)	211.3	Xing et al. (2021)
GOCA beads	Pb(II), Hg(II), Cd(II)	602, 374, 181	Arshad et al. (2019)
GO/PAMAMs	Pb(II), Cd(II), Cu(II), Mn(II)	568.18, 253.81, 68.68, 18.29	Zhang et al. (2014)
CS/GO-SH	Cu(II), Pb(II), Cd(II)	425, 447, 177	Li et al. (2015)
MMSP-GO	Pb(II), Cd(II)	333, 167	Wang et al. (2013)
PVK–GO	Pb(II)	887.98	Musico et al. (2013)
$MnFe_2O_4$/GO	Pb(II), As(III), As(V)	673, 146, 207	Kumar et al. (2014)
EDTA-mGO	Pb(II), Hg(II), Cu(II)	508.4, 268.4, 301.2	Cui et al. (2015)
GO/L-Trp	Cu(II), Pb(II)	588, 222	Tan et al. (2015)
PAH-GO	Cu(II)	349.04	Xing et al. (2015)
GO–CD–PPY NC	Cr(VI)	666.67	Chauke et al. (2015)
RGO/NiO	Cr(VI)	198	Zhang et al. (2018)
PAS–GO	U(VI), Eu(III)	310.63, 243.90	Zhao et al. (2017)
Chitosan/GO	Pb(II), Cu(II), Cr(VI)	461.3, 423.8, 310.4	Najafabadi et al. (2015)
GO	As(III)	1432.80	"Adsorption of Arsenic on Graphene Oxide, Reduced Graphene Oxide, and their Fe_3O_4 Doped Nanocomposites" (2021)
GO	Au(III)	149.20	Liu et al. (2012)
GO	Cd(II)	83.80	Sitko et al. (2013)
GO	Co(II)	21.28	Lingamdinne et al. (2016)
GO	Cr(VI)	1.22	Mondal and Chakraborty (2020)

(continued)

Table 3.1 (continued)

Material	Heavy metal ions	Adsorption tendency (mg/g)	References
GO	Cu(II)	294.00	Sitko et al. (2013)
GO	Eu(III)	175.44	Sun et al. (2012)
GO	Hg(II)	8.65	Tene et al. (2022)
GO	Mn(II)	32.00	Leiva et al. (2021)
GO	Ni(II)	62.30	Tan et al. (2015)
GO	Pb(II)	1119.00	Sitko et al. (2013)
GO	Sb(III)	8.06	Leng et al. (2012)
GO	Sr(II)	5.93	Zhao et al. (2016)
GO	Th(IV)	58.59	Xu et al. (2016)
GO	U(VI)	299.00	Li et al. (2012)
GO	Zn(II)	345.00	Sitko et al. (2013)

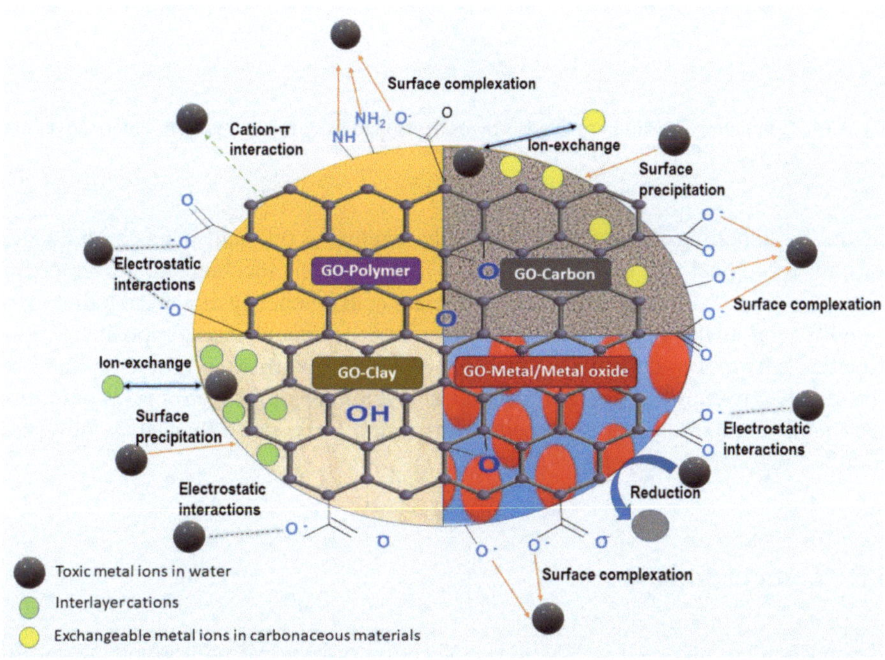

Fig. 3.13 Illustration of possible underlying adsorption mechanisms of graphene-based composites for metal ions (Bhushan et al. 2025)

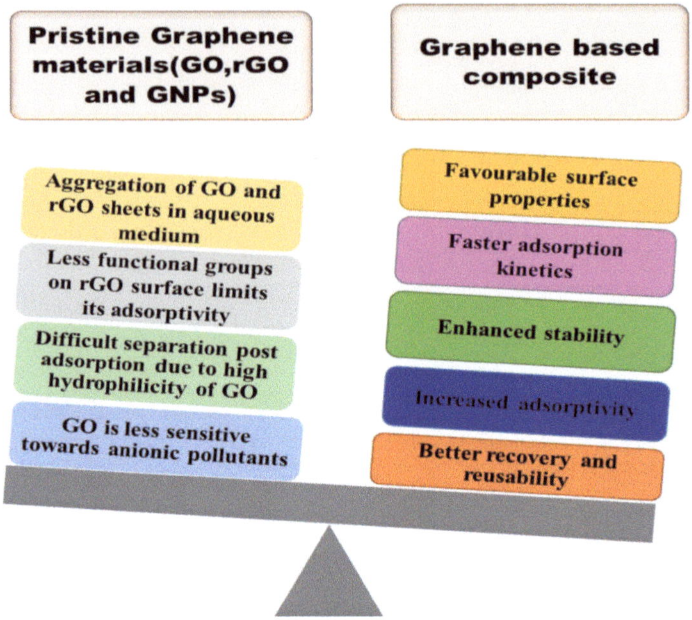

Fig. 3.14 Comparative evaluation of adsorbents developed from pristine graphene derivatives and graphene-derived composites (Bhushan et al. 2025)

factors for labelling an adsorbent to be the exclusive one and it has significance for both ecological and economic aspects. The synthesis mechanisms for graphene derivatives involve strong oxidising agents and acids; hence, using them judiciously is highly vital. In this regard, almost all the graphene-derived nanocomposites exhibit excellent adsorption–desorption attributes and highlight marked efficiency for HM ions removal even after several regeneration cycles. The average removal output of maximum graphene-based composites was found to be more than 80%, observed after 3–5 regeneration cycles.

3.5 Conclusions

The mechanism involved in HM ions adsorption or removal with graphene and its derivatives, though mentioned in the previous chapter, however, optimisation of the adsorption efficiency of graphene NMS has been described in the current chapter. Various drawbacks linked with the adsorption attributes of pristine graphene and its derivatives have been seen to be overcome by developing nanocomposites of graphene NMS. These nanocomposites are synthesised through various techniques, including polymerisation, melt blending, exfoliation, adsorption, and emulsion-based mechanisms. The adsorption kinetics obtained for different HM ions using diverse graphene-based nanocomposites indicate extremely high adsorption efficiency of these nanocomposites as compared to their constituent materials. Based on the current chapter, it is thus concluded that nanocomposites of graphene play a key role in the removal of HM ions effectively, and they exhibit superior adsorption characteristics along with fast kinetics, enhanced stability, and outstanding reusability and regenerability, compared to pristine graphene and its derivatives.

References

"Adsorption of arsenic on graphene oxide, reduced graphene oxide, and their Fe_3O_4 doped nanocomposites." Biointerface Res Appl Chem 12(5):6196–6210. https://doi.org/10.33263/BRIAC125.61966210

Abaszadeh M, Hosseinzadeh R, Tajbakhsh M, Ghasemi S (2022) The synthesis of functionalized magnetic graphene oxide with 5-amino-1,10-phenanthroline and investigation of its dual application in C-N coupling reactions and adsorption of heavy metal ions. J Mol Struct 1261:132832. https://doi.org/10.1016/j.molstruc.2022.132832

Adak NC, Chhetri S, Kim NH, Murmu NC, Samanta P, Kuila T (2018) Static and dynamic mechanical properties of graphene oxide-incorporated woven carbon fiber/epoxy composite. J Mater Eng Perform 27(3):1138–1147. https://doi.org/10.1007/s11665-018-3201-5

Alexandre M, Dubois P (2000) Polymer-layered silicate nanocomposites: Preparation, properties and uses of a new class of materials. Mater Sci Eng R Rep 28(1–2):1–63. https://doi.org/10.1016/S0927-796X(00)00012-7

Arshad F, Selvaraj M, Zain J, Banat F, Haija MA (2019) Polyethylenimine modified graphene oxide hydrogel composite as an efficient adsorbent for heavy metal ions. Sep Purif Technol 209:870–880. https://doi.org/10.1016/j.seppur.2018.06.035

Atif R, Shyha I, Inam F (2016) Mechanical, thermal, and electrical properties of graphene-epoxy nanocomposites—a review. Polymers (Basel) 8(8):281. https://doi.org/10.3390/polym8080281

Avouris P, Xia F (2012) Graphene applications in electronics and photonics. MRS Bull 37(12):1225–1234. https://doi.org/10.1557/mrs.2012.206

Ayala-Claveria M, Carlesi C, Puig J, Olguin G (2024) The impact of adsorption property modification by crosslinkers on graphene oxide membrane separation performance. Processes 12(11):2320. https://doi.org/10.3390/pr12112320

Bao C, Song L, Xing W, Yuan B, Wilkie CA, Huang J, Guo Y, Hu Y (2012) Preparation of graphene by pressurized oxidation and multiplex reduction and its polymer nanocomposites by masterbatch-based melt blending. J Mater Chem 22(13):6088. https://doi.org/10.1039/c2jm16203b

Bhushan B, Negi P, Nayak A, Goyal S (2025) Graphene composites for water remediation: An overview of their advanced performance with focus on challenges and future prospects. Adv Compos Hybrid Mater 8(1):55. https://doi.org/10.1007/s42114-024-01088-x

Bustero I, Gaztelumendi I, Obieta I, Mendizabal MA, Zurutuza A, Ortega A, Alonso B (2020) Free-standing graphene films embedded in epoxy resin with enhanced thermal properties. Adv Compos Hybrid Mater 3(1):31–40. https://doi.org/10.1007/s42114-020-00136-6

Cai Y, Fadil Y, Jasinski F, Thickett SC, Agarwal V, Zetterlund PB (2019) Miniemulsion polymerization using graphene oxide as surfactant: In situ grafting of polymers. Carbon N Y 149:445–451. https://doi.org/10.1016/j.carbon.2019.04.063

Cao Y, Zhang H, Song F, Huang T, Ji J, Zhong Q, Chu W, Xu Q (2018) UiO-66-NH_2/GO composite: Synthesis, characterization and Co_2 adsorption performance. Materials 11(4):589. https://doi.org/10.3390/ma11040589

Chauke VP, Maity A, Chetty A (2015) High-performance towards removal of toxic hexavalent chromium from aqueous solution using graphene oxide-alpha cyclodextrin-polypyrrole nanocomposites. J Mol Liq 211:71–77. https://doi.org/10.1016/j.molliq.2015.06.044

Chen L, Li Y, Hu S, Sun J, Du Q, Yang X, Ji Q, Wang Z, Wang D, Xia Y (2016) Removal of methylene blue from water by cellulose/graphene oxide fibres. J Exp Nanosci 11(14):1156–1170. https://doi.org/10.1080/17458080.2016.1198499

Chi Z, Zhu Y, Liu W, Huang H, Li H (2021) Selective removal of As(III) using magnetic graphene oxide ion-imprinted polymer in porous media: Potential effect of external magnetic field. J Environ Chem Eng 9(4):105671. https://doi.org/10.1016/j.jece.2021.105671

Cui L, Wang Y, Gao L, Hu L, Yan L, Wei Q, Du B (2015) EDTA functionalized magnetic graphene oxide for removal of Pb(II), Hg(II) and Cu(II) in water treatment: Adsorption mechanism and separation property. Chem Eng J 281:1–10. https://doi.org/10.1016/j.cej.2015.06.043

Das TK, Sakthivel TS, Jeyaranjan A, Seal S, Bezbaruah AN (2020) Ultra-high arsenic adsorption by graphene oxide iron nanohybrid: Removal mechanisms and potential applications. Chemosphere 253:126702. https://doi.org/10.1016/j.chemosphere.2020.126702

Fadil Y, Agarwal V, Jasinski F, Thickett SC, Minami H, Zetterlund PB (2019) Electrically conductive polymer/rGO nanocomposite films at ambient temperature via miniemulsion polymerization using GO as surfactant. Nanoscale 11(14):6566–6570. https://doi.org/10.1039/C9NR00816K

Fadil Y, Thickett SC, Agarwal V, Zetterlund PB (2022) Synthesis of graphene-based polymeric nanocomposites using emulsion techniques. Prog Polym Sci 125:101476. https://doi.org/10.1016/j.progpolymsci.2021.101476

Fan L, Luo C, Sun M, Li X, Qiu H (2013) Highly selective adsorption of lead ions by water-dispersible magnetic chitosan/graphene oxide composites. Colloids Surf B Biointerfaces 103:523–529. https://doi.org/10.1016/j.colsurfb.2012.11.006

Farajvand M, Grazioso V, García-Violini D, Ringwood JV (2023) Uncertainty estimation in wave energy systems with applications in robust energy maximising control. Renew Energy 203:194–204. https://doi.org/10.1016/j.renene.2022.12.054

Feng Z-Q, Yuan X, Wang T (2020) Porous polyacrylonitrile/graphene oxide nanofibers designed for high efficient adsorption of chromium ions (VI) in aqueous solution. Chem Eng J 392:123730. https://doi.org/10.1016/j.cej.2019.123730

Fu S, Sun Z, Huang P, Li Y, Hu N (2019) Some basic aspects of polymer nanocomposites: A critical review. Nano Materials Science 1(1):2–30. https://doi.org/10.1016/j.nanoms.2019.02.006

Gonçalves G, Domingues EM, Ferreira N, Ranawadia K, Henriques B, Bessa A, Tavares D, Martins N, Pereira E, Marques PAAP (2024) Enhanced Hg(II) removal using thiourea-functionalized graphene oxide: Lab to pilot scale evaluation. Sep Purif Technol 351:128053. https://doi.org/10.1016/j.seppur.2024.128053

Han X, Liu Y, Xiong L, Huang H, Zhang Q, Li L, Yu X, Wei L (2019) Facile assembly of polyaniline/graphene oxide composite hydrogels as adsorbent for Cr(VI) removal. Polym Compos 40(S2). https://doi.org/10.1002/pc.25161

Hosseinzadeh H, Ramin S (2018) Effective removal of copper from aqueous solutions by modified magnetic chitosan/graphene oxide nanocomposites. Int J Biol Macromol 113:859–868. https://doi.org/10.1016/j.ijbiomac.2018.03.028

Ibrahim A, Klopocinska A, Horvat K, Abdel Hamid Z (2021) Graphene-based nanocomposites: Synthesis, mechanical properties, and characterizations. Polymers (Basel) 13(17):2869. https://doi.org/10.3390/polym13172869

Jaleh B, Khalilipour A, Habibi S, Niyaifar M, Nasrollahzadeh M (2017) Synthesis, characterization, magnetic and catalytic properties of graphene oxide/Fe_3O_4. J Mater Sci: Mater Electron 28(6):4974–4983. https://doi.org/10.1007/s10854-016-6151-4

Kerisit S, Rosso KM (2005) Charge transfer in FeO: A combined molecular-dynamics and *ab initio* study. J Chem Phys 123(22). https://doi.org/10.1063/1.2137319

Khatun F, Aziz AA, Sim LC (2020) Preparation of Reduced Graphene Oxide (RGO) modified titanium dioxide nanotube (TNTs) as visible light effective catalyst for the conversion of CO_2

to CH_4. IOP Conf Ser Mater Sci Eng 736(4):042002. https://doi.org/10.1088/1757-899X/736/4/042002

Kim H, Miura Y, Macosko CW (2010) Graphene/polyurethane nanocomposites for improved gas barrier and electrical conductivity. Chem Mater 22(11):3441–3450. https://doi.org/10.1021/cm100477v

Krishna RH, Chandraprabha MN, Samrat K, Krishna Murthy TP, Manjunatha C, Kumar SG (2023) Carbon nanotubes and graphene-based materials for adsorptive removal of metal ions—A review on surface functionalization and related adsorption mechanism. Appl Surf Sci Adv 16:100431. https://doi.org/10.1016/j.apsadv.2023.100431

Kumar ASK, Jiang S-J (2016) Chitosan-functionalized graphene oxide: A novel adsorbent an efficient adsorption of arsenic from aqueous solution. J Environ Chem Eng 4(2):1698–1713. https://doi.org/10.1016/j.jece.2016.02.035

Kumar S, Nair RR, Pillai PB, Gupta SN, Iyengar MAR, Sood AK (2014) Graphene oxide–$MnFe_2O_4$ magnetic nanohybrids for efficient removal of lead and arsenic from water. ACS Appl Mater Interfaces 6(20):17426–17436. https://doi.org/10.1021/am504826q

Leiva E, Tapia C, Rodríguez C (2021) Removal of Mn(II) from acidic wastewaters using graphene oxide–ZnO nanocomposites. Molecules 26(9):2713. https://doi.org/10.3390/molecules26092713

Leng Y, Guo W, Su S, Yi C, Xing L (2012) Removal of Antimony(III) from aqueous solution by graphene as an adsorbent. Chem Eng J 211–212:406–411. https://doi.org/10.1016/j.cej.2012.09.078

Li Z, Chen F, Yuan L, Liu Y, Zhao Y, Chai Z, Shi W (2012) Uranium(VI) adsorption on graphene oxide nanosheets from aqueous solutions. Chem Eng J 210:539–546. https://doi.org/10.1016/j.cej.2012.09.030

Li X, Zhou H, Wu W, Wei S, Xu Y, Kuang Y (2015) Studies of heavy metal ion adsorption on Chitosan/Sulfydryl-functionalized graphene oxide composites. J Colloid Interface Sci 448:389–397. https://doi.org/10.1016/j.jcis.2015.02.039

Li S, Li Y, Han X, Zhao X, Zhao Y (2019) High-efficiency enhancement on thermal and electrical properties of epoxy nanocomposites with core-shell carbon foam template-coated graphene. Compos Part A Appl Sci Manuf 120:95–105. https://doi.org/10.1016/j.compositesa.2019.01.032

Lin KA, Yang H, Hsu F (2018) Zr-metal organic framework and derivatives for adsorptive and photocatalytic removal of acid dyes. Water Environ Res 90(2):144–154. https://doi.org/10.2175/106143017X15054988926604

Ling L, Zhang W (2014) Sequestration of arsenate in zero-valent iron nanoparticles: Visualization of intraparticle reactions at angstrom resolution. Environ Sci Technol Lett 1(7):305–309. https://doi.org/10.1021/ez5001512

Lingamdinne LP, Koduru JR, Roh H, Choi Y-L, Chang Y-Y, Yang J-K (2016) Adsorption removal of Co(II) from waste-water using graphene oxide. Hydrometallurgy 165:90–96. https://doi.org/10.1016/j.hydromet.2015.10.021

Liu L, Li C, Bao C, Jia Q, Xiao P, Liu X, Zhang Q (2012) Preparation and characterization of chitosan/graphene oxide composites for the adsorption of Au(III) and Pd(II). Talanta 93:350–357. https://doi.org/10.1016/j.talanta.2012.02.051

Lu B, Yuan X, Ren Y, Shi Q, Wang S, Dong J, Nan Z (2018) Cost-effective three dimensional Ag/polymer dyes/graphene-carbon spheres hybrids for high performance nonenzymatic sensor and its application in living cell H_2O_2 detection. Bioelectrochemistry 123:103–111. https://doi.org/10.1016/j.bioelechem.2018.05.001

Luo S, Xu X, Zhou G, Liu C, Tang Y, Liu Y (2014) Amino siloxane oligomer-linked graphene oxide as an efficient adsorbent for removal of Pb(II) from wastewater. J Hazard Mater 274:145–155. https://doi.org/10.1016/j.jhazmat.2014.03.062

Melitas N, Wang J, Conklin M, O'Day P, Farrell J (2002) Understanding soluble arsenate removal kinetics by zerovalent iron media. Environ Sci Technol 36(9):2074–2081. https://doi.org/10.1021/es011250y

Mondal NK, Chakraborty S (2020) Adsorption of Cr(VI) from aqueous solution on graphene oxide (GO) prepared from graphite: equilibrium, kinetic and thermodynamic studies. Appl Water Sci 10(2):61. https://doi.org/10.1007/s13201-020-1142-2
Musico YLF, Santos CM, Dalida MLP, Rodrigues DF (2013) Improved removal of lead(ii) from water using a polymer-based graphene oxide nanocomposite. J Mater Chem A Mater 1(11):3789. https://doi.org/10.1039/c3ta01616a
Najafabadi HH, Irani M, Rad LR, Sojoudi A, Haririan I (2015) Correction: Removal of Cu 2+, Pb 2+ and Cr 6+ from aqueous solutions using a chitosan/graphene oxide composite nanofibrous adsorbent. RSC Adv 5(29):22390–22390. https://doi.org/10.1039/C5RA90015H
Neolaka YAB, Lawa Y, Naat JN, Riwu AAP, Iqbal M, Darmokoesoemo H, Kusuma HS (2020) The adsorption of Cr(VI) from water samples using graphene oxide-magnetic (GO-Fe_3O_4) synthesized from natural cellulose-based graphite (kusambi wood or Schleichera oleosa): Study of kinetics, isotherms and thermodynamics. J Market Res 9(3):6544–6556. https://doi.org/10.1016/j.jmrt.2020.04.040
Nie Z-J, Guo Q-F, Xia H, Song M-M, Qiu Z-J, Fan S-T, Chen Z-H, Zhang S-X, Zhang S, Li B-J (2021) Cyclodextrin self-assembled graphene oxide aerogel microspheres as broad-spectrum adsorbent for removing dyes and organic micropollutants from water. J Environ Chem Eng 9(2):104749. https://doi.org/10.1016/j.jece.2020.104749
Parida K, Mishra KG, Dash SK (2012) Adsorption of Copper(II) on NH 2 -MCM-41 and its application for epoxidation of styrene. Ind Eng Chem Res 51(5):2235–2246. https://doi.org/10.1021/ie200109h
Patole AS, Patole SP, Kang H, Yoo J-B, Kim T-H, Ahn J-H (2010) A facile approach to the fabrication of graphene/polystyrene nanocomposite by in situ microemulsion polymerization. J Colloid Interface Sci 350(2):530–537. https://doi.org/10.1016/j.jcis.2010.01.035
Paul DR, Robeson LM (2008) Polymer nanotechnology: Nanocomposites. Polymer (Guildf) 49(15):3187–3204. https://doi.org/10.1016/j.polymer.2008.04.017
Rafiee MA, Rafiee J, Wang Z, Song H, Yu Z-Z, Koratkar N (2009) Enhanced mechanical properties of nanocomposites at low graphene content. ACS Nano 3(12):3884–3890. https://doi.org/10.1021/nn9010472
Ravi A, Punith Kumar MK, Rekha MY, Santosh MS, Srivastava C (2020) Graphene based nanocomposites: Synthesis, properties and application as electrochemical sensors, pp 1–20. https://doi.org/10.1016/bs.coac.2020.08.004
Sammaiah A, Huang W, Wang X (2018) Synthesis of magnetic Fe_3O_4/graphene oxide nanocomposites and their tribological properties under magnetic field. Mater Res Express 5(10):105006. https://doi.org/10.1088/2053-1591/aadaab
Sasaki K, Nakano H, Wilopo W, Miura Y, Hirajima T (2009) Sorption and speciation of arsenic by zero-valent iron. Colloids Surf A Physicochem Eng Asp 347(1–3):8–17. https://doi.org/10.1016/j.colsurfa.2008.10.033
Sengupta R, Bhattacharya M, Bandyopadhyay S, Bhowmick AK (2011) A review on the mechanical and electrical properties of graphite and modified graphite reinforced polymer composites. Prog Polym Sci 36(5):638–670. https://doi.org/10.1016/j.progpolymsci.2010.11.003
Sheikhmohammadi A, Mohseni SM, Khodadadi R, Sardar M, Abtahi M, Mahdavi S, Keramati H, Dahaghin Z, Rezaei S, Almasian M, Sarkhosh M, Faraji M, Nazari S (2017) Application of graphene oxide modified with 8-hydroxyquinoline for the adsorption of Cr (VI) from wastewater: Optimization, kinetic, thermodynamic and equilibrium studies. J Mol Liq 233:75–88. https://doi.org/10.1016/j.molliq.2017.02.101
Shen M-Y, Chang T-Y, Hsieh T-H, Li Y-L, Chiang C-L, Yang H, Yip M-C (2013) Mechanical properties and tensile fatigue of graphene nanoplatelets reinforced polymer nanocomposites. J Nanomater 2013(1). https://doi.org/10.1155/2013/565401
Singh S, Naik TSSK, BU, Khan NA, Wani AB, Behera SK, Nath B, Bhati S, Singh J, Ramamurthy PC (2022) A systematic study of arsenic adsorption and removal from aqueous environments using novel graphene oxide functionalized UiO-66-NDC nanocomposites. Sci Rep 12(1):15802. https://doi.org/10.1038/s41598-022-18959-2

Sitko R, Turek E, Zawisza B, Malicka E, Talik E, Heimann J, Gagor A, Feist B, Wrzalik R (2013) Adsorption of divalent metal ions from aqueous solutions using graphene oxide. Dalton Trans 42(16):5682. https://doi.org/10.1039/c3dt33097d

Sun Y, Wang Q, Chen C, Tan X, Wang X (2012) Interaction between Eu(III) and graphene oxide nanosheets investigated by batch and extended x-ray absorption fine structure spectroscopy and by modeling techniques. Environ Sci Technol 46(11):6020–6027. https://doi.org/10.1021/es300720f

Sun Q, Wang L, Li Y, Li L, Li S, Zhu G (2023) Highly efficient removal of mercury ions from aqueous solutions by thiol-functionalized graphene oxide. Water (Basel) 15(14):2529. https://doi.org/10.3390/w15142529

Tan P, Sun J, Hu Y, Fang Z, Bi Q, Chen Y, Cheng J (2015) Adsorption of Cu2+, Cd2+ and Ni2+ from aqueous single metal solutions on graphene oxide membranes. J Hazard Mater 297:251–260. https://doi.org/10.1016/j.jhazmat.2015.04.068

Tan M, Liu X, Li W, Li H (2015) Enhancing sorption capacities for Copper(II) and Lead(II) under weakly acidic conditions by <scp>l</scp>-Tryptophan-functionalized graphene oxide. J Chem Eng Data 60(5):1469–1475. https://doi.org/10.1021/acs.jced.5b00015

Taylor P (1998) Ostwald ripening in emulsions. Adv Colloid Interface Sci 75(2):107–163. https://doi.org/10.1016/S0001-8686(98)00035-9

Tene T, Bellucci S, Guevara M, Arias Arias F, Sáez Paguay MÁ, Quispillo Moyota JM, Arias Polanco M, Scarcello A, Vacacela Gomez C, Straface S, Caputi LS, Torres FJ (2022) Adsorption of mercury on oxidized graphenes. Nanomaterials, 12(17):3025. https://doi.org/10.3390/nano12173025

Thy LTM, Thuong NH, Tu TH, Nam HM, Hieu NH, Phong MT (2019) Synthesis of magnetic iron oxide/graphene oxide nanocomposites for removal of cadmium ions from water. Adv Nat Sci: Nanosci Nanotechnol 10(2):025006. https://doi.org/10.1088/2043-6254/ab1b79

Tuček J, Prucek R, Kolařík J, Zoppellaro G, Petr M, Filip J, Sharma VK, Zbořil R (2017) Zero-valent iron nanoparticles reduce Arsenites and arsenates to As(0) firmly embedded in core-shell superstructure: Challenging strategy of arsenic treatment under anoxic conditions. ACS Sustain Chem Eng 5(4):3027–3038. https://doi.org/10.1021/acssuschemeng.6b02698

Uhl FM, Wilkie CA (2002) Polystyrene/graphite nanocomposites: Effect on thermal stability. Polym Degrad Stab 76(1):111–122. https://doi.org/10.1016/S0141-3910(02)00003-4

Uysal Unalan I, Cerri G, Marcuzzo E, Cozzolino CA, Farris S (2014) Nanocomposite films and coatings using inorganic nanobuilding blocks (NBB): Current applications and future opportunities in the food packaging sector. RSC Adv 4(56):29393–29428. https://doi.org/10.1039/C4RA01778A

Vazquez A, Cyras VP, Alvarez VA, Moran JI (2012) Starch/clay nano-biocomposites, pp 287–321. https://doi.org/10.1007/978-1-4471-4108-2_11

Vo LQ, Vu A-T, Le TD, Huynh CD, Tran HV (2024) Fe_3O_4/Graphene oxide/chitosan nanocomposite: A smart nanosorbent for Lead(II) ion removal from contaminated water. ACS Omega. https://doi.org/10.1021/acsomega.4c00486

Wakamatsu J, Kawasaki M, Zetterlund PB, Okubo M (2007) Nitroxide-mediated radical polymerization in microemulsion. Macromol Rapid Commun 28(24):2346–2353. https://doi.org/10.1002/marc.200700576

Wang J, Chen B (2015) Adsorption and coadsorption of organic pollutants and a heavy metal by graphene oxide and reduced graphene materials. Chem Eng J 281:379–388. https://doi.org/10.1016/j.cej.2015.06.102

Wang Y, Liang S, Chen B, Guo F, Yu S, Tang Y (2013) Synergistic removal of Pb(II), Cd(II) and humic acid by Fe_3O_4@mesoporous silica-graphene oxide composites. PLoS ONE 8(6):e65634. https://doi.org/10.1371/journal.pone.0065634

Wang H, Liu Y, Zeng G, Hu X, Hu X, Li T, Li H, Wang Y, Jiang L (2014) Grafting of β-cyclodextrin to magnetic graphene oxide via ethylenediamine and application for Cr(VI) removal. Carbohydr Polym 113:166–173. https://doi.org/10.1016/j.carbpol.2014.07.014

Wang Y, Lin K, Liu Y, Deng X (2022) Nanocomposites of functionalized metal−organic frameworks and magnetic graphene oxide for selective adsorption and efficient determination of Lead(II). J Solid State Chem 313:123300. https://doi.org/10.1016/j.jssc.2022.123300

Witomska S, Liu Z, Czepa W, Aliprandi A, Pakulski D, Pawluć P, Ciesielski A, Samorì P (2019) Graphene oxide hybrid with sulfur-nitrogen polymer for high-performance pseudocapacitors. J Am Chem Soc 141(1):482–487. https://doi.org/10.1021/jacs.8b11181

Wu C, Tu J, Liu W, Zhang J, Chu S, Lu G, Lin Z, Dang Z (2017) The double influence mechanism of pH on arsenic removal by nano zero valent iron: Electrostatic interactions and the corrosion of Fe0. Environ Sci Nano 4(7):1544–1552. https://doi.org/10.1039/C7EN00240H

Xing HT, Chen JH, Sun X, Huang YH, Su ZB, Hu SR, Weng W, Li SX, Guo HX, Wu WB, He YS, Li FM, Huang Y (2015) NH2-rich polymer/graphene oxide use as a novel adsorbent for removal of Cu(II) from aqueous solution. Chem Eng J 263:280–289. https://doi.org/10.1016/j.cej.2014.10.111

Xing C, Xia A, Yu L, Dong L, Hao Y, Qi X (2021) Enhanced Removal of Pb(II) from aqueous solution using EDTA–modified magnetic graphene oxide. Clean (Weinh), 49(4). https://doi.org/10.1002/clen.202000272

Xu H, Li G, Li J, Chen C, Ren X (2016) Interaction of Th(IV) with graphene oxides: Batch experiments, XPS investigation, and modeling. J Mol Liq 213:58–68. https://doi.org/10.1016/j.molliq.2015.11.022

Xu X, Zeng J, Wu Y, Wang Q, Wu S, Gu H (2022) Preparation and application of graphene-based materials for heavy metal removal in tobacco industry: A review. Separations 9(12):401. https://doi.org/10.3390/separations9120401

Xue X, Xu J, Baig SA, Xu X (2016) Synthesis of graphene oxide nanosheets for the removal of Cd(II) ions from acidic aqueous solutions. J Taiwan Inst Chem Eng 59:365–372. https://doi.org/10.1016/j.jtice.2015.08.019

Yan W, Vasic R, Frenkel AI, Koel BE (2012) Intraparticle reduction of Arsenite (As(III)) by nanoscale zerovalent Iron (nZVI) investigated with in situ x-ray absorption spectroscopy. Environ Sci Technol 46(13):7018–7026. https://doi.org/10.1021/es2039695

Yang Y, Wang W, Li M, Wang H, Zhao M, Wang C (2018) Preparation of PANI grafted at the edge of graphene oxide sheets and its adsorption of Pb(II) and methylene blue. Polym Compos 39(5):1663–1673. https://doi.org/10.1002/pc.24114

Yang J, Hou B, Wang J, Tian B, Bi J, Wang N, Li X, Huang X (2019) Nanomaterials for the removal of heavy metals from wastewater. Nanomaterials 9(3):424. https://doi.org/10.3390/nano9030424

Yang S, Li Q, Chen L, Chen Z, Pu Z, Wang H, Yu S, Hu B, Chen J, Wang X (2019) Ultrahigh sorption and reduction of Cr(VI) by two novel core-shell composites combined with Fe_3O_4 and MoS_2. J Hazard Mater 379:120797. https://doi.org/10.1016/j.jhazmat.2019.120797

Yang S, Li Q, Chen L, Chen Z, Hu B, Wang H, Wang X (2020) Synergistic removal and reduction of U(VI) and Cr(VI) by Fe_3S_4 micro-crystal. Chem Eng J 385:123909. https://doi.org/10.1016/j.cej.2019.123909

Yuan J-M, Fan Z-F, Chen X-H, Chen X-H, Wu Z-J, He L-P (2009) Preparation of polystyrene–multi-walled carbon nanotube composites with individual-dispersed nanotubes and strong interfacial adhesion. Polymer (Guildf) 50(14):3285–3291. https://doi.org/10.1016/j.polymer.2009.04.065

Zetterlund PB, Wakamatsu J, Okubo M (2009) Nitroxide-mediated radical polymerization of styrene in aqueous microemulsion: Initiator efficiency, compartmentalization, and nitroxide phase transfer. Macromolecules 42(18):6944–6952. https://doi.org/10.1021/ma9012829

Zhang H-B, Zheng W-G, Yan Q, Jiang Z-G, Yu Z-Z (2012) The effect of surface chemistry of graphene on rheological and electrical properties of polymethylmethacrylate composites. Carbon N Y 50(14):5117–5125. https://doi.org/10.1016/j.carbon.2012.06.052

Zhang F, Wang B, He S, Man R (2014) Preparation of graphene-oxide/polyamidoamine dendrimers and their adsorption properties toward some heavy metal ions. J Chem Eng Data 59(5):1719–1726. https://doi.org/10.1021/je500219e

Zhang L, Luo H, Liu P, Fang W, Geng J (2016) A novel modified graphene oxide/chitosan composite used as an adsorbent for Cr(VI) in aqueous solutions. Int J Biol Macromol 87:586–596. https://doi.org/10.1016/j.ijbiomac.2016.03.027

Zhang K, Li H, Xu X, Yu H (2018) Synthesis of reduced graphene oxide/NiO nanocomposites for the removal of Cr(VI) from aqueous water by adsorption. Microporous Mesoporous Mater 255:7–14. https://doi.org/10.1016/j.micromeso.2017.07.037

Zhao X, Zhang Q, Chen D, Lu P (2010) Enhanced mechanical properties of graphene-based poly(vinyl alcohol) composites. Macromolecules 43(5):2357–2363. https://doi.org/10.1021/ma902862u

Zhao Y, Guo C, Fang H, Jiang J (2016) Competitive adsorption of Sr(II) and U(VI) on graphene oxide investigated by batch and modeling techniques. J Mol Liq 222:263–267. https://doi.org/10.1016/j.molliq.2016.07.032

Zhao D, Chen L, Xu M, Feng S, Ding Y, Wakeel M, Alharbi NS, Chen C (2017) Amino siloxane oligomer modified graphene oxide composite for the efficient capture of U(VI) and Eu(III) from aqueous solution. ACS Sustain Chem Eng 5(11):10290–10297. https://doi.org/10.1021/acssuschemeng.7b02316

Zhou L, Zhang G, Tian J, Wang D, Cai D, Wu Z (2017) Functionalized Fe_3O_4 @C nanospheres with adjustable structure for efficient hexavalent chromium removal. ACS Sustain Chem Eng 5(11):11042–11050. https://doi.org/10.1021/acssuschemeng.7b02983

Chapter 4
Challenges and Future Directions in Graphene-Based Heavy Metal Remediation

4.1 Challenges and Suggested Measures

Synthesis and economic challenges: Graphene can be synthesised through various top-down techniques such as plasma etching, liquid-phase exfoliation, laser ablation, electrochemical exfoliation, chemical reduction, and ball milling as well as bottom-up techniques including epitaxial growth, thermal pyrolysis, and chemical vapor deposition (CVD) as already elaborated in the Chap. 1. Though bottom-up techniques are capable of developing extremely precise graphene nanostructures as well as graphene sheets exhibiting enlarged surface area (Ruffieux et al. 2016; Lopes et al. 2018). However, graphene synthesis for industrial-scale applications needs to be embellished to attain a high-quality product. Besides, reproducibility and uniformity of graphene flakes are another important aspect, as well as production rate and scalability, which are major hurdles in already established synthesis techniques such as CVD. Moreover, inconsistent quality obtained via different producers calls for standardising the synthesis and grading process for graphene production (Lin et al. 2019).

It is observed that a few graphene synthesis techniques, such as laser ablation, play a major role in the mass production of graphene. The as-synthesised graphene material exhibits superior quality with the fewest defects. Due to a facile and quick synthesis procedure, the laser ablation technique represents a cost-effective way of producing industrial-scale graphene. Another process, viz., the electrochemical synthesis process for graphene, possesses great potential in developing high-quality graphene material with tunable oxidation level without the usage of hazardous chemicals, as it involves electrochemical activation. Besides, the purification methods involved in this process are very simple, unlike other purifying processes (Yu et al. 2015).

Technical challenges: Though graphene NMS, such as GO, rGO, as well as functionalized graphene, have been used extensively for adsorptive removal of environmental pollutants, including HMIS, however, various technical challenges need

P. Mahajan and V. Khanna, *Graphene in the Environment*, Environmental Science and Engineering, https://doi.org/10.1007/978-3-031-96708-5_4

to be addressed for achieving precision in this matter on a commercial scale. These challenges include exploring the efficacy of graphene-based adsorbents in complex and dynamic environmental substrates rather than focusing on the output achieved in a stable and non-complicated polluted system. This area needs a lot of effort, as most of the studies in the literature have focused on the application of graphene-derived adsorbents on a single adsorbent system, though in real-world applications, the environment to be remediated represents a lot of complexity and dynamism. Further, it is observed from various research data that graphene-based materials as adsorbents pose a lot of challenges in their recovery and reusability (Kong et al. 2021). If this happens repeatedly, these graphene NMS will get accumulated excessively and reside in the natural environment for the time unlimited. Consequently, the quality of the environment will be degraded, and it will induce a compromising situation for healthy living; thus, rather than acting as a pollutant remediator, it may itself become a threat to a healthy and safe environment. Another major technical hurdle is poor knowledge about the interaction mechanism existing between the graphene-based adsorbents and the environmental pollutants. This is because there exists a huge difference in the physicochemical properties of different graphene-based adsorbents. Although numerous reports can be extracted from the literature regarding the application of graphene-based NMS for HMIS adsorption, a standardised evaluation process is lacking to assess the tendency of these adsorbents' affinity towards a few HMIS. Moreover, exploring the interaction mechanism for each metal ion through various experimentation is a time-consuming as well as highly laborious task. Few studies have inferred that there exists a linear correlation between the characteristics of GO and HMIS (Kong et al. 2020). Currently, research data indicating the correlation between the adsorptive output of graphene-based materials and characteristic parameters of HMIS is highly deficient. Hence, working on this aspect may help in analysing the ability of graphene-derived adsorbents in adsorption of unidentified HMIS, thereby saving time, resources, and manpower.

Another technical challenge is linked to the chemical modification of graphene NMS for optimisation of their adsorption ability for HMIS. Pristine graphene is often functionalized through chemical modifications using functional groups containing O, S, and N. However, no research has been conducted regarding the chelation tendency of these functional groups towards different HMS. If the theoretical data regarding the chelation ability of different functional groups towards different HMS can be obtained, a suitable guide for graphene-based adsorbents with outstanding adsorption efficiency and low-cost production can be developed through modifications with appropriate functional groups for a particular metal. Besides, the other technical challenge is the co-existence of HMIS with organic pollutants, which leads to the formation of HM-organic compounds, particularly in the wastewater. The reports regarding the adsorption of HMIS co-existing with organic pollutants using graphene-based adsorbents are lacking. Maximum research is conducted towards single pollutant systems, however, the adsorption mechanism for HMIS in the presence of organic pollutants is different. Hence, it is necessary to elaborate on the behaviour, either competitive or synergistic, of HMIS towards graphene-based adsorbents when co-existing with organic pollutants (Kong et al. 2021).

Scalability and sustainability challenges: The bottom-up synthesis mechanisms for graphene put a hefty cost on production, whereas top-down techniques involve the usage of obnoxious chemicals. Hence, scalability and sustainability an important issues that need to be addressed. For this, highly productive synthesis mechanisms with bearable cost and least environmental impact must be explored and implemented for optimisation of the HM remediation process using graphene NMS. Besides, the fabrication of functionalized graphene NMS is a key perspective in obtaining highly efficient adsorbents for HMI removal from the ecosystem. However, the methodologies involved in the modification and functionalization of graphene NMS generally exhibit high execution cost, rendering limited implication of graphene-derived NMS in practical applications (Xu et al. 2022). It is believed that using biomass precursors for fabricating graphene and its composites may provide cost-effective and eco-friendly synthesis mechanisms.

Environmental challenges: The ph of the substrate, particularly aqueous media, has a significant impact on the adsorption efficiency of graphene NMS and its composites. In addition to inducing surface charge on the adsorbent, ph also converts the speciation of metal ions based on pollutants. The fact that there is a positive charge on the adsorbent's surface when aqueous ph < pH_{pzc} and a negative charge when ph > pH_{pzc} is known very well. Besides, a study reported a significant increase in the adsorption of Cr ions by GO-PANI composite till ph 6, which declined with further increase in ph. The declined adsorptive capacity of this composite for Pb ions is due to the conversion of Cr ions into hydroxylated complexes in response to ph beyond 6 (Han et al. 2019). In general, it is observed that the adsorption of metal ions by GO-based composites exhibited high dependency on the ph of the medium, acidic ph being the favourable condition. Whereas, dye removal is feasible under a wide ph range. Therefore, ph optimisation is also a major challenge in attaining the precision for adsorptive removal of HMIS using graphene-based adsorbents. Besides, a study reported the positive impact of co-existing metal ions such as Cd, Ni, Pb, Cu, and Zn in the adsorptive removal of Hg ions through cyclodextrin/graphene composite (CD@RGO). The underlying mechanism involves the creation of cation-induced novel adsorption sites on the surface of the composite material (Ema et al. 2017). Contrary to this, a study reported a decline efficiency of GO composites in the adsorption of particular metal ions in the co-existence of other metal ions. Therefore, developing highly analytical systems regarding the contributory environmental factors for accelerating the adsorption of HMIS using graphene NMS is highly demanded.

Toxicity challenges: As graphene-based NMS are gaining immense importance for adsorptive removal of HMIS from the environment, hence its rapid production, characterisation, and application have excessively increased the environmental and human exposure. Various studies regarding the long-term and short-term in vivo toxicity and eco-distribution of graphene NMS have been conducted by different researchers (Ema et al. 2017; Guo and Mei 2014). Owing to intense biological and chemical activities, graphene-based NMS are highly capable in generation of generating reactive oxygen species (ROS) in living organisms. The toxicity of graphene NMS is due to its extremely small size and large surface area, and it varies with its surface attributes, functionalized groups, number of layers, concentration, cell type, duration, penetration pathway, and temperature. Excessive exposure to graphene NMS induces oxidative stress, physical damage, inflammatory reactions, destruction to DNA and cell membranes, necrosis, chromosomal deformities, and cell death. Besides, these NMS exhibit huge toxicity in the plants, including cabbage, red spinach, and tomatoes (Ali et al. 2019).

4.2 Conclusions

The alarmingly increasing rate of HMS accumulation in the environment needs to be addressed in a highly efficient and sustainable way due to its life-threatening impact on the biota. Nanotechnology and nanomaterials are gaining immense importance in diverse fields, including environmental applications. Amongst various NMS, a 2-dimensional NM, viz., graphene NM, exhibit unique physicochemical properties such as extremely elevated specific surface area, outstanding electronic conductivity, robust mechanical strength, great flexibility for chemical modifications, and functionalization for developing nanocomposites. Graphene NMS exhibit higher efficiency in the adsorption of pollutants as compared to conventional carbon-based adsorbents. Moreover, on comparison with pristine graphene, the oxygen functionalities of graphene derivatives such as GO and rGO further elevate the adsorption attributes by providing ample active sites. Besides, there are several synthesis mechanisms for graphene, including both bottom-up and top-down approaches. Most of these techniques are easy to operate and involve an appropriate cost, further adding value to the graphene-based environmental remediation process. The basic mechanism for heavy metal removal by graphene NMS involves the process of adsorption facilitated by hydrophobic interaction, precipitation, ion-exchange, πstacking, surface complexation, and electrostatic interaction. The pristine graphene and its derivatives, though applied widely in the field of environmental remediation, lack efficiency as expected. Hence, focused analysis has been conducted to recognise the structural and functional deficiencies in graphene-based adsorbents. This induced a park in the material scientists to optimise the physicochemical attributes of graphene NMS, which led to the formation of nanocomposites. Nanocomposites refer to composite materials whose one of the constituents is at the nanometer scale and the other one can be any 2D material. Graphene-based nanocomposites can be developed through the process of polymerisation, melt blending, exfoliation, adsorption, and emulsion-based mechanisms. On characterisation and implication of the graphene-based nanocomposites in the field of adsorptive removal of HMIS from the environment, it is inferred that they are enriched with physicochemical properties and exhibit much superiority as compared to graphene NMS. Therefore, these nanocomposites are capable of overcoming the drawbacks linked with the adsorption attributes of pristine graphene and its derivatives. It is thus estimated that graphene-derived nanocomposites are crucial in the removal of HMIS effectively, and they exhibit superior adsorption characteristics along with fast kinetics, enhanced stability, and outstanding reusability and regenerability, compared to pristine graphene and its derivatives. Nevertheless, attaining precision is a never-ending task in any technology, and various challenges keep coming with the passage of process and time. Besides developing nanocomposites for enhancing the efficiency of graphene-based adsorbents, the challenges in other aspects are still unresolved, such as the secondary impact on the health and environment of this remediation process, which involves toxic chemicals in various fabrication mechanisms of graphene NMS. Besides, over-consumption of these NMS will ultimately lead to their accumulation in the environmental substrates, if not disposed of properly leading to various toxic impacts on the biota. Moreover, the chemical modification and functionalization techniques put an immense economic burden. In addition, various technical challenges such as the performance of graphene-based adsorbents in the removal of particular metal ions in the presence of various other ions, as well as varying ph of the aqueous media, persist. The scalability and sustainability issues in using graphene NMS for adsorptive removal of HMIS should also be tackled to optimise this remediation process.

4.3 Future Perspectives

Integration of advanced and innovative techniques with nanotechnology may open up a new era of environmental remediation. AI-directed NM's designing may have a robust impact on their adsorption and catalytic efficiency towards toxic environmental pollutants, including HMIS. The efficiency of nano adsorbents and nano catalysts in environmental remediation is already well known and exhibits higher efficiency than the customary techniques (66). However, developing nanoparticles and nanorobots guided by AI may achieve higher precision by targeting and degrading particular pollutants, such as HMS and organic pollutants. Besides, analysing the long-term health and environmental impacts of NM's accumulation is a major challenge which can be addressed via AI. AI-based simulation techniques and advanced predictive modelling through AI algorithms can analyse complete datasets involving materials attributes, pollutant characteristics and behaviour, and environmental parameters. This may lead to early projection of ecological impacts due to NMs applications. Moreover, it may also discern the complex interactions existing between the NMS and different ecosystems, facilitating well-planned decision-making towards the mitigation of associated environmental risks. In addition, AI may aid the optimisation of NM's attributes as well as the remediation process through automation in designing and broadcasting the NMS, owing to superior pollutant remediation abilities with the least environmental impact. In addition to AI, machine learning (ML) algorithms are of significant importance in analysing experimental as well as simulation data for predicting maximal material compositions, designing, and fabrication techniques highly impactful for pollutant removal applications. AI-based optimisation algorithms are also useful in fine-tuning the remediation processes, including deployment of NM's, and environmental quality assessing protocols, which can significantly enhance the remediation output with minimal resource utilisation and environmental disturbance. Hence, AI and ML integration along with nanotechnology-based

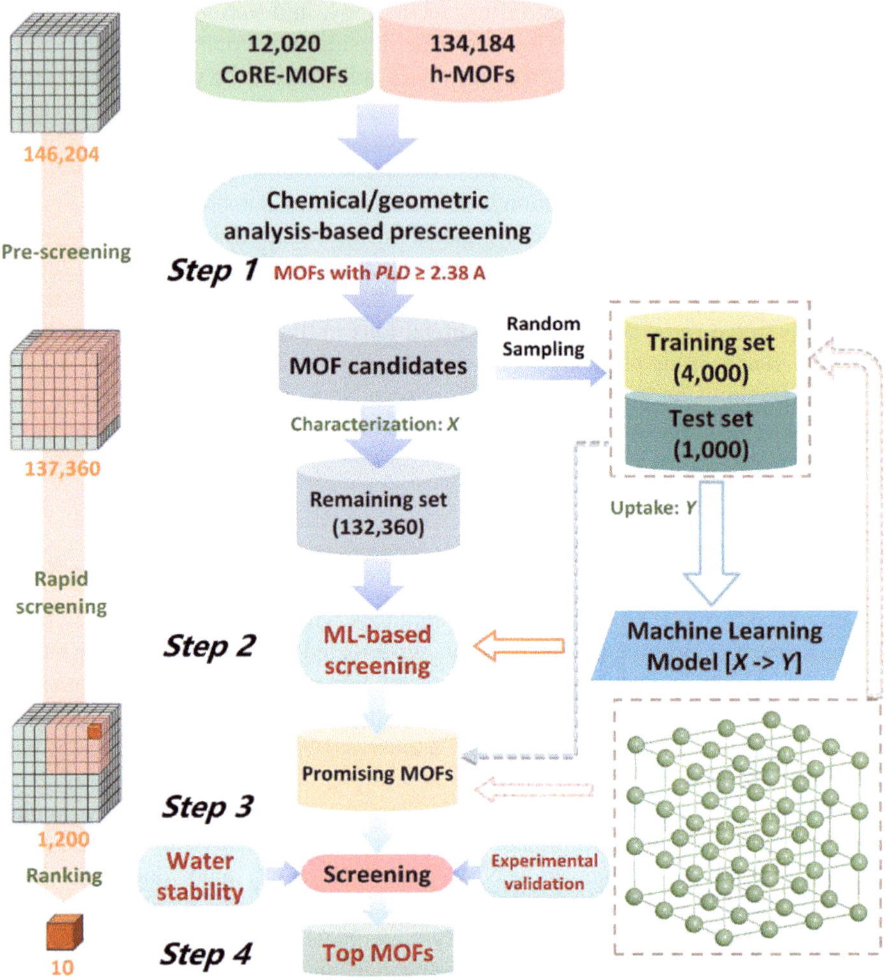

Fig. 4.1 Represents machine learning (ML) technique which assists in quick screening of Pb ions adsorption efficiency in an aqueous media (Yuan et al. 2024)

remediation processes can significantly employ the strategy of developing environmental eco-friendly techniques for HMI removal, thereby accentuating the criteria of cost-effectiveness, ecological sustainability, and scalability (Fig. 4.1).

References

Ali I, Basheer AA, Mbianda XY, Burakov A, Galunin E, Burakova I, Mkrtchyan E, Tkachev A, Grachev V (2019) Graphene based adsorbents for remediation of noxious pollutants from wastewater. Environ Int 127:160–180. https://doi.org/10.1016/j.envint.2019.03.029

Ema M, Gamo M, Honda K (2017) A review of toxicity studies on graphene-based nanomaterials in laboratory animals. Regul Toxicol Pharmacol 85:7–24. https://doi.org/10.1016/j.yrtph.2017.01.011

Guo X, Mei N (2014) Assessment of the toxic potential of graphene family nanomaterials. J Food Drug Anal 22(1):105–115. https://doi.org/10.1016/j.jfda.2014.01.009

Han X, Liu Y, Xiong L, Huang H, Zhang Q, Li L, Yu X, Wei L (2019) Facile assembly of polyaniline/graphene oxide composite hydrogels as adsorbent for Cr(VI) removal. Polym Compos 40(S2). https://doi.org/10.1002/pc.25161

Kong Q, Preis S, Li L, Luo P, Wei C, Li Z, Hu Y, Wei C (2020) Relations between metal ion characteristics and adsorption performance of graphene oxide: A comprehensive experimental and theoretical study. Sep Purif Technol 232:115956. https://doi.org/10.1016/j.seppur.2019.115956

Kong Q, Shi X, Ma W, Zhang F, Yu T, Zhao F, Zhao D, Wei C (2021) Strategies to improve the adsorption properties of graphene-based adsorbent towards heavy metal ions and their compound pollutants: A review. J Hazard Mater 415:125690. https://doi.org/10.1016/j.jhazmat.2021.125690

Lalwani G, D'Agati M, Khan AM, Sitharaman B (2016) Toxicology of graphene-based nanomaterials. Adv Drug Deliv Rev 105:109–144. https://doi.org/10.1016/j.addr.2016.04.028

Lin L, Peng H, Liu Z (2019) Synthesis challenges for graphene industry. Nat Mater 18(6):520–524. https://doi.org/10.1038/s41563-019-0341-4

Lopes LC, da Silva LC, Vaz BG, Oliveira ARM, Oliveira MM, Rocco MLM, Orth ES, Zarbin AJG (2018) Facile room temperature synthesis of large graphene sheets from simple molecules. Chem Sci 9(37):7297–7303. https://doi.org/10.1039/C8SC02818D

Ruffieux P, Wang S, Yang B, Sánchez-Sánchez C, Liu J, Dienel T, Talirz L, Shinde P, Pignedoli CA, Passerone D, Dumslaff T, Feng X, Müllen K, Fasel R (2016) On-surface synthesis of graphene nanoribbons with zigzag edge topology. Nature 531(7595):489–492. https://doi.org/10.1038/nature17151

Sanchez VC, Jachak A, Hurt RH, Kane AB (2012) Biological interactions of graphene-family nanomaterials: An interdisciplinary review. Chem Res Toxicol 25(1):15–34. https://doi.org/10.1021/tx200339h

Xu X, Zeng J, Wu Y, Wang Q, Wu S, Gu H (2022) Preparation and application of graphene-based materials for heavy metal removal in tobacco industry: A review. Separations 9(12):401. https://doi.org/10.3390/separations9120401

Yu P, Lowe SE, Simon GP, Zhong YL (2015) Electrochemical exfoliation of graphite and production of functional graphene. Curr Opin Colloid Interface Sci 20(5–6):329–338. https://doi.org/10.1016/j.cocis.2015.10.007

Yuan L, Xu M, Zhang Y, Gao Z, Zhang L, Cheng C, Ji C, Hua M, Lv L, Zhang W (2024) Machine learning-assisted screening of metal-organic frameworks (MOFs) for the removal of heavy metals in aqueous solution. Sep Purif Technol 339:126732. https://doi.org/10.1016/j.seppur.2024.126732

MIX
Papier aus verantwortungsvollen Quellen
Paper from responsible sources
FSC® C105338

If you have any concerns about our products,
you can contact us on
ProductSafety@springernature.com

In case Publisher is established outside the EU,
the EU authorized representative is:
Springer Nature Customer Service Center GmbH
Europaplatz 3, 69115 Heidelberg, Germany

Printed by Libri Plureos GmbH
in Hamburg, Germany